D1523691

Simulated Annealing:
Theory and Applications

Mathematics and Its Applications

Simulated Annealing: Theory and Applications

by

P. J. M. van Laarhoven

and

E. H. L. Aarts

Philips Research Laboratories,
Eindhoven, The Netherlands

Kluwer Academic Publishers
Dordrecht / Boston / London

Library of Congress Cataloging in Publication Data

Laarhoven, Peter J. M. van.
Simulated annealing.

(Mathematics and its applications)
Bibliography: p.
Includes index.
1. Combinatorial optimization. 2. Algorithms. I. Aarts, Emile H. L. II. Title. III. Series: Mathematics and its applications (D. Reidel Publishing Company)
QA402.5.L3 1987 511 87–9666
ISBN 90–277–2513–6

Published by Kluwer Academic Publishers,
P.O. Box 17, 3300 AA Dordrecht, The Netherlands.

Kluwer Academic Publishers incorporates
the publishing programmes of
D. Reidel, Martinus Nijhoff, Dr W. Junk and MTP Press.

Sold and distributed in the U.S.A. and Canada
by Kluwer Academic Publishers,
101 Philip Drive, Norwell, MA 02061, U.S.A.

In all other countries, sold and distributed
by Kluwer Academic Publishers Group,
P.O. Box 322, 3300 AH Dordrecht, The Netherlands.

Reprinted with corrections 1988

Reprinted 1989

3-1289-400 ts

Printed in The Netherlands

SERIES EDITOR'S PREFACE

Approach your problems from the right end and begin with the answers. Then one day, perhaps you will find the final question.

'The Hermit Clad in Crane Feathers' in R. van Gulik's *The Chinese Maze Murders*.

It isn't that they can't see the solution. It is that they can't see the problem.

G.K. Chesterton. *The Scandal of Father Brown* 'The point of a Pin'.

Growing specialization and diversification have brought a host of monographs and textbooks on increasingly specialized topics. However, the "tree" of knowledge of mathematics and related fields does not grow only by putting forth new branches. It also happens, quite often in fact, that branches which were thought to be completely disparate are suddenly seen to be related.

Further, the kind and level of sophistication of mathematics applied in various sciences has changed drastically in recent years: measure theory is used (non-trivially) in regional and theoretical economics; algebraic geometry interacts with physics; the Minkowsky lemma, coding theory and the structure of water meet one another in packing and covering theory; quantum fields, crystal defects and mathematical programming profit from homotopy theory; Lie algebras are relevant to filtering; and prediction and electrical engineering can use Stein spaces. And in addition to this there are such new emerging subdisciplines as "experimental mathematics", "CFD", "completely integrable systems", "chaos, synergetics and large-scale order", which are almost impossible to fit into the existing classification schemes. They draw upon widely different sections of mathematics. This programme, Mathematics and Its Applications, is devoted to new emerging (sub)disciplines and to such (new) interrelations as exempla gratia:

- a central concept which plays an important role in several different mathematical and/or scientific specialized areas;
- new applications of the results and ideas from one area of scientific endeavour into another;
- influences which the results, problems and concepts of one field of enquiry have and have had on the development of another.

The Mathematics and Its Applications programme tries to make available a careful selection of books which fit the philosophy outlined above. With such books, which are stimulating rather than definitive, intriguing rather than encyclopaedic, we hope to contribute something towards better communication among the practitioners in diversified fields.

Simulated annealing is a perfect example of how introducing ideas from a completely different and at first sight, totally unrelated field may yield large and unexpected benefits. In this case, a completely new and powerful tool for global (combinatorial) optimization.

Annealing is the process of heating a solid and cooling it slowly so as to remove strain and crystal imperfections. During this process the free energy of the solid is minimized. The initial heating is necessary to avoid becoming trapped in a local minimum. Virtually every function can be viewed as the free energy of some system and thus studying and imitating how nature reaches a minimum during the annealing process should yield optimization algorithms. Just what imitation, i.e. simulation, here means mathematically is of course thoroughly explained in the book and so is the underlying relation with (statistical) physics (cf. Chapter 4 for the latter). The result is the tool called "simulated annealing", which, since its inception in 1982, has become a remarkably powerful tool in global optimization.

Nature (and therefore technology) is, of course, full of minimizing processes, another being Maupertuis' least action principle and more tools in optimization may well result from other imitations of Nature's ways of achieving things with least effort. As a matter of fact the algorithms in B.S. Razumikhin's book "Physical models and equilibrium methods in programming and economics" (Reidel 1984) are also based on such ideas.

As stated above, simulated annealing as an optimization tool started in 1982, developed quickly and vigorously, and by now it seems to have reached a certain maturity, witness the present volume, the first monograph on the topic. This does not mean that there are no large fascinating problems. For instance, the tool is no panacea and also very little seems to be known about a priori conditions on optimization problems which say something about how well simulated annealing is likely to perform. Another question is what "cooling speed" is optimal. For these and other aspects one needs experience and expert guidance. This book provides the latter.

The unreasonable effectiveness of mathematics in science ...

Eugene Wigner

Well, if you know of a better 'ole, go to it.

Bruce Bairnsfather

What is now proved was once only imagined.

William Blake

As long as algebra and geometry proceeded along separate paths, their advance was slow and their applications limited.

But when these sciences joined company they drew from each other fresh vitality and thenceforward marched on at a rapid pace towards perfection.

Joseph Louis Lagrange.

Bussum, March 1987

Michiel Hazewinkel

Contents

Preface

1 Introduction **1**

2 Simulated annealing **7**
2.1 Introduction of the algorithm 7
2.2 Mathematical model of the algorithm 12

3 Asymptotic convergence results **17**
3.1 The homogeneous algorithm 17
3.1.1 Introduction 17
3.1.2 Existence of the stationary distribution 18
3.1.3 Convergence of the stationary distribution . . . 20
3.2 The inhomogeneous algorithm 27
3.2.1 Introduction 27
3.2.2 Sufficient conditions for convergence 28
3.2.3 Necessary and sufficient conditions for convergence 35

4 The relation with statistical physics **39**
4.1 Introduction 39
4.2 Equilibrium dynamics 40
4.3 Typical behaviour of the simulated annealing algorithm 44
4.4 Phase transitions 48
4.5 Relation with spin glasses 50

5 Towards implementing the algorithm **55**
5.1 Introduction 55
5.2 Conceptually simple cooling schedules 59

5.3 More elaborate cooling schedules 62
5.4 Improvements of the generation mechanism for transitions 71

6 Performance of the simulated annealing algorithm 77
6.1 Introduction 77
6.2 Worst-case performance results 79
6.3 Probabilistic value analysis 82
6.4 Performance on combinatorial optimization problems . 88

7 Applications 99
7.1 Introduction 99
7.2 Applications in computer-aided circuit design 100
7.2.1 Introduction 100
7.2.2 Placement 101
7.2.3 Routing 110
7.2.4 Other applications in computer-aided circuit design 118
7.3 Applications in other areas 128
7.3.1 Image processing 128
7.3.2 Boltzmann machines 131
7.3.3 Miscellaneous applications 135

8 Some miscellaneous topics 139
8.1 Parallel implementations 139
8.1.1 General algorithms 140
8.1.2 Tailored algorithms 143
8.1.3 Special parallel implementations 147
8.2 Continuous optimization 148
8.2.1 Generalization of simulated annealing to continuous problems 148
8.2.2 Applications of simulated annealing to continuous problems 151

9 Summary and conclusions 153

Bibliography 157

Index 176

Preface

As early as 1953, Metropolis *et al.* [MET53] proposed an algorithm for the efficient simulation of the evolution of a solid to thermal equilibrium. It took almost thirty years[1] before Kirkpatrick *et al.* [KIR82] and, independently, Černy [ČER85] realized that there exists a profound analogy between minimizing the cost function of a combinatorial optimization problem and the slow cooling of a solid until it reaches its low energy ground state and that the optimization process can be realized by applying the Metropolis criterion. By substituting cost for energy and by executing the Metropolis algorithm at a sequence of slowly decreasing temperature values Kirkpatrick and his co-workers obtained a combinatorial optimization algorithm, which they called 'simulated annealing'. Since then, the research into this algorithm and its applications has evolved into a field of study in its own.

The authors have witnessed the maturation of this field at close quarters: at Philips Research Laboratories we have been involved in research on the theory and applications of simulated annealing since 1983. About a year ago, we concluded that the theoretical basis of the algorithm had reached a certain level of saturation and that major contributions were to be expected predominantly with respect to new applications. It was for this reason that we decided the time was ripe to write a review of the theory and applications of simulated annealing. Time has indeed proved us right: in 1986, we counted some dozens of contributions on applications of simulated annealing

[1] Two earlier papers on optimization of continuous functions, [PIN70] and [KHA81], in which the analogy between statistical mechanics and optimization is already noticed, have remained widely overlooked.

and virtually no major theoretical contributions. Moreover, the past few years a few succinct reviews and overviews appeared: [HAJ85], [BOU86], [WIL86a], [AAR87a] and [AAR87b] (the last two were based on a preliminary version of this monograph), indicating a need for review papers.

The aim of this monograph is to present a thorough treatment of the theoretical basis of the algorithm, to discuss computational experience with the algorithm and to review applications of the algorithm in various disciplines. With respect to the last item, we do not even pretend to list completely all existing applications: the area of applications is all but open-ended and new ones are reported almost monthly. However, we do aim to give an indication of the large variety of applications.

While writing a monograph on simulated annealing, a technique which has been successfully applied in so many different disciplines, we were reminded of the following anecdote:[2]

Once upon a time a fire broke out in a hotel, where just then a scientific conference was held. It was night and all guests were sound asleep. As it happened, the conference was attended by researchers from a variety of disciplines. The first to be awakened by the smoke was a mathematician. His first reaction was to run immediately to the bathroom, where, seeing that there was still water running from the tap, he exclaimed: "There is a solution!". At the same time, however, the physicist went to see the fire, took a good look and went back to his room to get an amount of water, which would be just sufficient to extinguish the fire. The electronic engineer was not so choosy and started to throw buckets and buckets of water on the fire. Finally, when the biologist awoke, he said to himself: "The fittest will survive" and he went back to sleep.

In the spirit of this anecdote, we would like to recommend chapter 2 (asymptotic convergence) to mathematicians, chapters 4 and 5 (implementation aspects) to physicists, chapter 7 (problem solving) to electrical engineers, the complete monograph to combinatorial optimizers and another book to biologists.

A monograph like this could not have been written without the sup-

---

[2]This anecdote was originally told by dr. C.L. Liu at the Workshop on Statistical Physics in Engineering and Biology, held in Lech, Austria, in July 1986.

port of many. We are especially grateful to some of our colleagues, in particular Gerard Beenker and Guido Janssen, who have helped, by discussion and criticism, with the preparation of the manuscript, and to Theo Claasen and Eric van Utteren, who encouraged us to cast this review in the form of a monograph. We express our gratitude to Scott Kirkpatrick and to C.L. Liu for the permission to reproduce figures from references [KIR82], [LEO85a] and [VEC83]. We are grateful to David Johnson, Jan Karel Lenstra and the other (anonymous) referees for their suggestions. Finally, we would like to thank those who sent us (p)reprints of their work. Please, continue to do so!

Peter J.M. van Laarhoven　　Eindhoven, February 1987
Emile H.L. Aarts

NOTE TO THE REVISED PRINTING

The main changes in the reprinted edition are in the references, which we brought up to date. Apart from the correction of errata, the text has remained essentially unchanged.
Since this book was published in 1987, the flow of papers on simulated annealing and, especially, on its applications has not dimished in the slightest. Of these we would like to mention explicitly the books of Aarts and Korst [AAR88b] and Wong *et al.* [WON88] and the review article by Collins *et al.* [COL87].

Peter J.M. van Laarhoven　　Eindhoven, September 1988
Emile H.L. Aarts

Chapter 1

Introduction

During the past decades, the role of *optimization* has steadily increased in such diverse areas as, for example, electrical engineering, operations research, computer science and communication. *Linear and non-linear optimization* [WAG75], the search for an optimum of a function of *continuous* variables, has seen major breakthroughs in the fifties and sixties (the best known example being the *simplex algorithm* to solve *linear programming* problems, introduced by Dantzig in 1947 [DAN63]). Major results in *combinatorial optimization* [PAP82], the search for optima of functions of *discrete* variables, were obtained predominantly in the seventies. For example, for the *Travelling Salesman Problem* (TSP) [LAW85], probably the best known problem in combinatorial optimization, both approximation and optimization algorithms of high quality became available in the seventies ([LIN73] and [GRÖ77], respectively) and nowadays, TSPs of up to 318 cities are solved to optimality in reasonable computation times (less than 10 minutes CPU time on an IBM 370/168-computer) [CRO80].
However, even today, many large-scale combinatorial optimization problems can only be solved approximately on present-day computers, which is closely related to the fact that many of these problems have been proved *NP-hard* [GAR79] (or the corresponding decision problem *NP-complete*). Such problems are unlikely to be well solvable, i.e. solvable by an amount of computational effort which is bounded by a *polynomial function* of the size of the problem. An extensive classification of these problems is given in the well-known book of Garey

and Johnson [GAR79], where 320 NP-complete problems are cited. The distinction between NP-complete problems and those in P (the class of problems solvable in *polynomial time*) seems closely related to the distinction between hard and easy problems. Computational experience has increased evidence for this relation: though there is, of course, the possibility of such contrasts as an $\mathcal{O}(1.001^n)$ *exponential-time algorithm* and an $\mathcal{O}(n^{100})$ *polynomial-time algorithm*, these kinds of complexities hardly ever seem to occur in practice [COO83], [JOHN85a].

Thus, we are left in a situation, where in practice many large-scale combinatorial optimization problems cannot be solved to optimality, because the search for an optimum requires prohibitive amounts of computation time (e.g., optimal solutions for TSPs of more than 318 cities have not been published in the literature and the aforementioned 10 minutes of CPU-time only came after years of research for just **one** 318-city instance). One is forced to use *approximation algorithms* or *heuristics*, for which there is usually no guarantee that the solution found by the algorithm is optimal, but for which polynomial bounds on the computation time can be given. Thus, knowledge of the quality of the final solution is traded off against computation time.

An instance of a combinatorial optimization problem is formalized as a pair $(\mathcal{R}, C)$, where $\mathcal{R}$ is the finite - or possibly countably infinite - *set of configurations* (also called *configuration space*) and C a *cost function*, $C : \mathcal{R} \to \mathbf{R}$, which assigns a real number to each configuration. For convenience, it is assumed that C is defined such that the lower the value of C, the better (with respect to the optimization criteria) the corresponding configuration, which can be done without loss of generality. The problem now is to find a configuration for which C takes its minimum value, i.e. an (optimal) configuration i_0 satisfying

$$C_{opt} = C(i_0) = \min_{i \in \mathcal{R}} C(i), \tag{1.1}$$

where C_{opt} denotes the optimum (minimum) cost (for the TSP, for example, configurations correspond to tours of the salesman and the objective is to find the tour of shortest length).

As mentioned before, there are two possible approaches when trying to solve a combinatorial optimization problem. One can either use an *optimization algorithm*, yielding a globally optimal solution in a possibly prohibitive amount of computation time (for large NP-hard problems) or an approximation algorithm, yielding an approximate solution in an acceptable amount of computation time. Approximation algorithms can be divided into two categories: algorithms tailored to a specific problem (*tailored algorithms*), and *general algorithms* applicable to a wide variety of combinatorial optimization problems. To avoid the intrinsic disadvantage of the algorithms in the first category (their limited applicability due to the problem dependence) it is desirable to have a general approximation algorithm, able to obtain near-optimal solutions for a wide variety of combinatorial optimization problems. The simulated annealing algorithm can be viewed as such an algorithm: it is a general optimization technique for solving combinatorial optimization problems. The algorithm is based on *randomization techniques*. However, it also incorporates a number of aspects related to *iterative improvement* algorithms. Since these aspects play a major role in the understanding of the simulated annealing algorithm, we first elaborate on the iterative improvement technique.

The application of an iterative improvement algorithm presupposes the definition of configurations, a cost function and a *generation mechanism*, i.e. a simple prescription to generate a *transition* from a configuration to another one by a small perturbation. The generation mechanism defines a *neighbourhood* $\mathcal{R}_i$ for each configuration i, consisting of all configurations that can be reached from i in one transition. Iterative improvement is therefore also known as *neighbourhood search* or *local search* [PAP82]. The algorithm can now be formulated as follows. Starting off at a given configuration, a sequence of iterations is generated, each iteration consisting of a possible transition from the current configuration to a configuration selected from the neighbourhood of the current configuration. If this neighbouring configuration has a lower cost, the current configuration is replaced by this neighbour, otherwise another neighbour is selected and compared for its cost value. The algorithm terminates when a configuration is obtained whose cost is no worse than any of its neighbours.

The disadvantages of iterative improvement algorithms can be formulated as follows:

- by definition, iterative improvement algorithms terminate in a local minimum and there is generally no information as to the amount by which this local minimum deviates from a global minimum;

- the obtained local minimum depends on the initial configuration, for the choice of which generally no guidelines are available;

- in general, it is not possible to give an upper bound for the computation time (e.g., the worst-case complexity of the iterative improvement algorithm for TSP based on *Lin's 2-opt strategy* [LIN65] is an open problem [JOHN85b]).

It should be clear, however, that iterative improvement does have the advantage of being generally applicable: configurations, a cost function and a generation mechanism are usually easy to define. Besides, though upper bounds for computation times are missing, a single run of an iterative improvement algorithm (for **one** initial configuration) can **on the average** be executed in a small amount of computation time.
To avoid some of the aforementioned disadvantages, one might think of a number of alternative approaches:

1. execution of the algorithm for a large number of initial configurations, say N (e.g., uniformly distributed over the set of configurations $\mathcal{R}$) at the cost of an increase in computation time [RIN83]; for $N \to \infty$, such an algorithm finds a global minimum with probability 1, if only for the fact that a global minimum is encountered as an initial configuration with probability 1 as $N \to \infty$;

2. (refinement of 1.) use of information gained from previous runs of the algorithm to improve the choice of an initial configuration for the next run [BAU86] (this information relates to the structure of the set of configurations);

3. introduction of a more complex generation mechanism (or, equivalently, enlargement of the neighbourhoods), in order to

be able to 'jump out' of the local minima corresponding to the simple generation mechanism. To choose the more complex generation mechanism properly requires detailed knowledge of the problem itself;

4. acceptance of transitions which correspond to an **increase** in the cost function in a limited way (in an iterative improvement algorithm only transitions corresponding to a **decrease** in cost are accepted).

The second and third approaches being usually strongly problem-dependent (restricting their general applicability), one is left with the first or last alternative. The first approach is a traditional way to solve combinatorial optimization problems approximately (the aforementioned iterative improvement algorithm based on Lin's 2-opt strategy [LIN65] is a well-known example). An algorithm that follows the fourth approach was independently introduced by Kirkpatrick *et al.* [KIR82], [KIR83] and Černy [ČER85]. It is generally known as *simulated annealing*, due to the analogy with the simulation of the *annealing of solids* it is based upon, but it is also known as *Monte Carlo annealing* [JEP83], *statistical cooling* [STO85], [AAR85a], *probabilistic hill climbing* [ROM85], *stochastic relaxation* [GEM84] or *probabilistic exchange algorithm* [ROSS86]. Solutions, obtained by simulated annealing, do not depend on the initial configuration and have a cost usually close to the minimum cost. Furthermore, it is possible to give a polynomial upper bound for the computation time for some implementations of the algorithm [AAR85a], [LUN86]. Thus, simulated annealing can be viewed as an algorithm that does not exhibit the disadvantages of iterative improvement and remains as generally applicable as iterative improvement. However, the gain of general applicability is sometimes undone by the computational effort, since the simulated annealing algorithm is slower than iterative improvement algorithms.

The aim of this monograph is to present a state-of-the-art overview of the theory and applications of simulated annealing. We start with an introduction to the algorithm, followed by the formulation of the mathematical model of the algorithm in terms of *Markov chains*.

In chapter 3 it is shown that ***asymptotically*** the algorithm finds a global minimum with probability 1, by using results from the theory of Markov chains. Thus, these results relate to the performance of the simulated annealing algorithm viewed as an optimization algorithm. In any implementation, however, the simulated annealing algorithm is an approximation algorithm. This is because the asymptotic nature of the aforementioned results relates to asymptotic values of parameters governing the convergence of the algorithm. In any implementation of the algorithm, these asymptotic values have to be approximated. The analogy between combinatorial optimization and ***statistical mechanics***, on which the algorithm is based and which is discussed in chapter 4, can be helpful in making appropriate choices for these approximations. These choices are further discussed in chapter 5, together with some other implementation aspects.
The quality of the final solution obtained by the algorithm is revisited in chapter 6, where the finite-time behaviour is considered: given a parameter choice, the performance of the algorithm, in terms of both quality of solution and computation time, is discussed on the basis of a number of well-known combinatorial optimization problems.
The wide variety in problems for which simulated annealing can be used as an approximation algorithm is illustrated by the discussion of applications in chapter 7. The major part of this chapter is devoted to applications in ***computer-aided circuit design***, applications in other areas are only briefly discussed. Some miscellaneous topics are dealt with in chapter 8 and the monograph is ended with some conclusions and remarks.

Chapter 2

Simulated annealing

2.1 Introduction of the algorithm

In its original form [KIR82], [ČER85] the simulated annealing algorithm is based on the analogy between the simulation of the annealing of solids and the problem of solving large combinatorial optimization problems. For this reason the algorithm became known as "simulated annealing". In condensed matter physics, annealing denotes a physical process in which a solid in a *heat bath* is heated up by increasing the temperature of the heat bath to a maximum value at which all particles of the solid randomly arrange themselves in the liquid phase, followed by cooling through slowly lowering the temperature of the heat bath. In this way, all particles arrange themselves in the low energy ground state of a corresponding lattice, provided the maximum temperature is sufficiently high and the cooling is carried out sufficiently slowly. Starting off at the maximum value of the temperature, the cooling phase of the annealing process can be described as follows. At each temperature value T, the solid is allowed to reach *thermal equilibrium*, characterized by a probability of being in a state with energy E given by the *Boltzmann distribution*:

$$Pr\{\mathbf{E} = E\} = \frac{1}{Z(T)} \cdot \exp\left(-\frac{E}{k_B T}\right), \tag{2.1}$$

where $Z(T)$ is a normalization factor, known as the *partition function*, depending on the temperature T and k_B is the *Boltzmann con-*

stant. The factor $\exp\left(-\frac{E}{k_B T}\right)$ is known as the *Boltzmann factor*. As the temperature decreases, the Boltzmann distribution concentrates on the states with lowest energy and finally, when the temperature approaches zero, only the minimum energy states have a non-zero probability of occurrence. However, it is well known [KIR82] that if the cooling is too rapid, i.e. if the solid is not allowed to reach thermal equilibrium for each temperature value, defects can be 'frozen' into the solid and metastable amorphous structures can be reached rather than the low energy crystalline lattice structure. Furthermore, in a process known in condensed matter physics as *quenching*, the temperature of the heat bath is lowered instantaneously, which results again in a freezing of the particles in the solid into one of the metastable amorphous structures.

To simulate the evolution to thermal equilibrium of a solid for a fixed value of the temperature T, Metropolis *et al.* [MET53] proposed a *Monte Carlo method*, which generates sequences of states of the solid in the following way. Given the current state of the solid, characterized by the positions of its particles, a small, randomly generated, perturbation is applied by a small displacement of a randomly chosen particle. If the difference in energy, ΔE, between the current state and the slightly perturbed one is **negative**, i.e. if the perturbation results in a lower energy for the solid, then the process is continued with the new state. If $\Delta E \geq 0$, then the probability of acceptance of the perturbed state is given by $\exp(-\frac{\Delta E}{k_B T})$. This acceptance rule for new states is referred to as the *Metropolis criterion*. Following this criterion, the system eventually evolves into thermal equilibrium, i.e. after a large number of perturbations, using the aforementioned acceptance criterion, the probability distribution of the states approaches the Boltzmann distribution, given by eq. 2.1. In statistical mechanics this Monte Carlo method, which is known as the *Metropolis algorithm*, is a well-known method used to estimate averages or integrals by means of random sampling techniques; for review articles see e.g. [BAR76], [BIN78], [HAS70] and [WOOD68].

The Metropolis algorithm can also be used to generate sequences of configurations of a combinatorial optimization problem. In that case, the configurations assume the role of the states of a solid while the cost function C and the *control parameter* c take the roles of energy and temperature, respectively. The simulated annealing algorithm can now be viewed as a sequence of Metropolis algorithms evaluated at a sequence of decreasing values of the control parameter. It can thus be described as follows. Initially, the control parameter is given a high value and a sequence of configurations of the combinatorial optimization problem is generated as follows. As in the iterative improvement algorithm a generation mechanism is defined, so that, given a configuration i, another configuration j can be obtained by choosing at random an element from the neighbourhood of i. The latter corresponds to the small perturbation in the Metropolis algorithm. Let $\Delta C_{ij} = C(j) - C(i)$, then the probability for configuration j to be the next configuration in the sequence is given by 1, if $\Delta C_{ij} \leq 0$, and by $\exp(-\frac{\Delta C_{ij}}{c})$, if $\Delta C_{ij} > 0$ (the Metropolis criterion). Thus, there is a non-zero probability of continuing with a configuration with higher cost than the current configuration. This process is continued until equilibrium is reached, i.e. until the probability distribution of the configurations approaches the Boltzmann distribution, now given by

$$Pr\{\text{configuration} = i\} \stackrel{\text{def}}{=} q_i(c) = \frac{1}{Q(c)} \cdot \exp\left(-\frac{C(i)}{c}\right), \tag{2.2}$$

where $Q(c)$ is a normalization constant depending on the control parameter c, being the equivalent of the aforementioned partition function.

The control parameter is then lowered in steps, with the system being allowed to approach equilibrium for each step by generating a sequence of configurations in the previously described way. The algorithm is terminated for some small value of c, for which virtually no deteriorations are accepted anymore. The final 'frozen' configuration is then taken as the solution of the problem at hand.

Thus, as with iterative improvement, we have again a generally applicable approximation algorithm: once configurations, a cost function and a generation mechanism (or, equivalently, a *neighbourhood structure*) are defined, a combinatorial optimization problem can be solved along the lines given by the description of the simulated annealing algorithm in pseudo-PASCAL in figure 2.1. Note that the acceptance criterion is implemented by drawing random numbers from a uniform distribution on $[0, 1)$ and comparing these with $\exp(-\frac{\Delta C_{ij}}{c})$.

```
PROCEDURE SIMULATED ANNEALING
begin
   INITIALIZE;
   M := 0;
   repeat
      repeat
         PERTURB(config. i → config. j,ΔC_ij);
         if ΔC_ij ≤ 0 then accept else
            if exp(−ΔC_ij/c) > random[0,1) then accept;
         if accept then UPDATE(configuration j);
      until equilibrium is approached sufficiently closely;
      c_{M+1} := f(c_M);
      M := M + 1;
   until stop criterion = true (system is 'frozen');
end.
```

Figure 2.1: Description of the annealing algorithm in pseudo-PASCAL.

Comparing iterative improvement and simulated annealing, it is apparent that the situation where the control parameter in the simulated annealing algorithm is set to 0 corresponds to a version of iterative improvement (it is not iterative improvement *per se*, because in an iterative improvement approach the neighbouring configurations are not necessarily examined in a random order). In the analogy with condensed matter physics, this corresponds to the aforementioned quenching process. Conversely, simulated annealing is a generalization of iterative improvement in that it accepts, with non-zero but gradually decreasing probability, deteriorations in the cost function. It is not clear, however, whether it performs better than repeated application of iterative improvement (for a number of different initial configurations): both algorithms converge asymptotically to a glob-

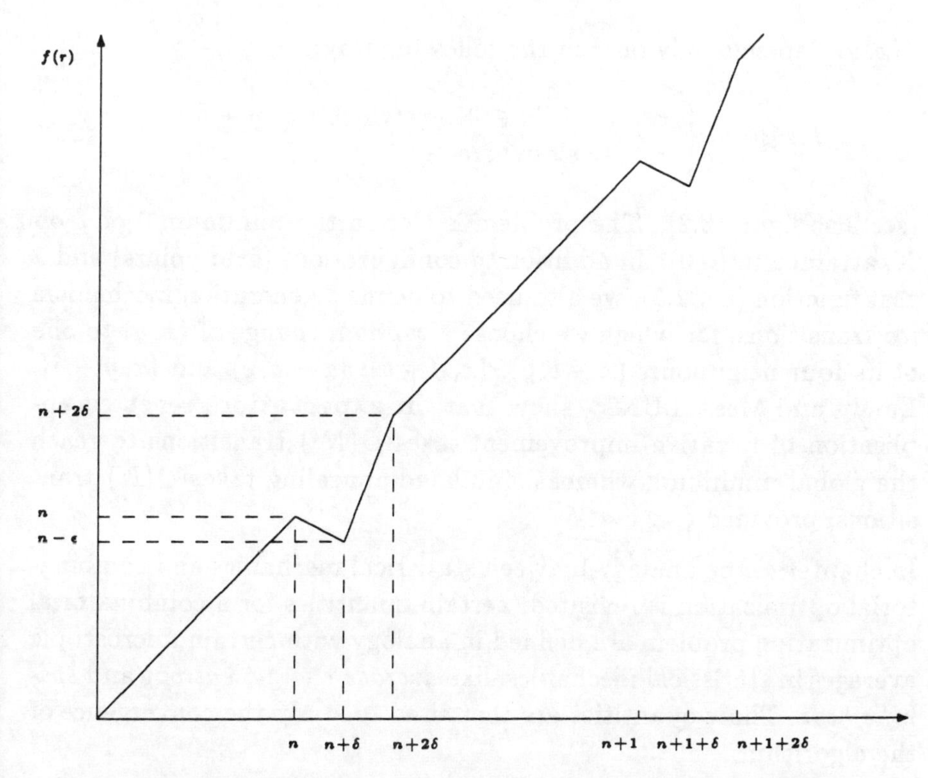

Figure 2.2: f as a function of the distance to the origin r.

ally minimal configuration of the problem at hand (for simulated annealing asymptotic convergence is proved in chapter 3; for repeated application of iterative improvement it is obvious that convergence is obtained for $N \to \infty$, where N is the number of initial configurations for which the algorithm is applied). However, to illustrate the difference in performance between both algorithms, we discuss the following example, taken from Lundy and Mees [LUN86], which consists of a problem for which simulated annealing indeed performs better than repeated application of iterative improvement. Consider the square $X = [-N, N] \times [-N, N] \subset \mathbf{R}^2$ (for some integer N) and let f be a function defined for each point $(x, y) \in X$, for which x and y are multiples of some positive rational number δ (in other words, f is defined on the points in X belonging to a regular grid with grid size δ). Let $r = \max\{|\,x\,|, |\,y\,|\}$ for each $(x, y) \in X$ and assume that

$f(x, y)$ depends only on r in the following way:

$$f(x,y) = \begin{cases} n - \epsilon & \text{if } \exists n \in \mathbf{N} \text{ for which } r = n + \delta \\ r & \text{elsewhere} \end{cases} \tag{2.3}$$

(see also figure 2.2). The problem is to find the minimum 0 of f on X, attained in (0,0). In addition to configurations (grid points) and a cost function (eq. 2.3), we also need to define a generation mechanism for transitions, for which we choose a random change of (x, y) to one of its four neighbours $(x + \delta, y)$, $(x, y + \delta)$, $(x - \delta, y)$ and $(x, y - \delta)$. Lundy and Mees [LUN86] show that, **in expectation**, repeated application of iterative improvement takes $O(N^2)$ transitions to reach the global minimum, whereas simulated annealing takes $O(N)$ transitions, provided $\epsilon \ll c \ll \delta$.

In chapter 4, the analogy between statistical mechanics and combinatorial optimization is revisited: certain quantities for a combinatorial optimization problem are defined in analogy with certain microscopic averages in statistical mechanics, like *average energy, entropy* and *specific heat*. These quantities are then used to study the convergence of the algorithm.

2.2 Mathematical model of the algorithm

Given a neighbourhood structure, simulated annealing can be viewed as an algorithm that continuously attempts to transform the current configuration into one of its neighbours. This mechanism is mathematically best described by means of a *Markov chain*: a sequence of trials, where the outcome of each trial depends only on the outcome of the previous one [FELL50]. In the case of simulated annealing, trials correspond to transitions and it is clear that the outcome of a transition depends only on the outcome of the previous one (i.e. the current configuration).

A Markov chain is described by means of a set of *conditional probabilities* $P_{ij}(k-1, k)$ for each pair of outcomes (i, j); $P_{ij}(k-1, k)$ is the probability that the outcome of the k-th trial is j, given that the outcome of the $k-1$-th trial is i. Let $a_i(k)$ denote the probability of outcome i at the k-trial, then $a_i(k)$ is obtained by solving the recursion:

$$a_i(k) = \sum_l a_l(k-1) \cdot P_{li}(k-1,k), \; k = 1, 2, \ldots, \tag{2.4}$$

where the sum is taken over all possible outcomes. Hereinafter, $\mathbf{X}(k)$ denotes the outcome of the k-th trial. Hence:

$$P_{ij}(k-1,k) = Pr\{\mathbf{X}(k) = j \mid \mathbf{X}(k-1) = i\} \tag{2.5}$$

and

$$a_i(k) = Pr\{\mathbf{X}(k) = i\}. \tag{2.6}$$

If the conditional probabilities do not depend on k, the corresponding Markov chain is called *homogeneous*, otherwise it is called *inhomogeneous*.

In the case of simulated annealing, the conditional probability $P_{ij}(k-1,k)$ denotes the probability that the k-th transition is a transition from configuration i to configuration j. Thus, $\mathbf{X}(k)$ is the configuration obtained after k transitions. In view of this, $P_{ij}(k-1,k)$ is called the *transition probability* and the $|\mathcal{R}| \times |\mathcal{R}|$-matrix $P(k-1,k)$ the *transition matrix*.

The transition probabilities depend on the value of the control parameter c, the analogue of the temperature in the physical annealing process. Thus, if c is kept constant, the corresponding Markov chain is homogeneous and its transition matrix $P = P(c)$ can be defined as:

$$P_{ij}(c) = \begin{cases} G_{ij}(c)A_{ij}(c) & \forall j \neq i \\ 1 - \sum_{l=1,l\neq i}^{|\mathcal{R}|} G_{il}(c)A_{il}(c) & j = i, \end{cases} \tag{2.7}$$

i.e. each transition probability is defined as the product of the following two conditional probabilities: the *generation probability* $G_{ij}(c)$ of generating configuration j from configuration i, and the *acceptance probability* $A_{ij}(c)$ of accepting configuration j, once it has been generated from i. The corresponding matrices $G(c)$ and $A(c)$ are called the *generation* and *acceptance matrix*, respectively. As a result of the definition in eq. 2.7, $P(c)$ is a *stochastic matrix*, i.e. $\forall i : \sum_j P_{ij}(c) = 1$.

Note that by defining the transition matrix $P(c)$ through eq. 2.7, we abstract from the original formulation of the algorithm [KIR82], [ČER85], where $G(c)$ is given by the *uniform distribution* on the neighbourhoods (transitions are implemented by choosing **at random** a neighbouring configuration j from the current configuration i) and $A(c)$ is given by the Metropolis criterion. Hereinafter, unless explicitly stated otherwise, we always refer with simulated annealing to the general class of algorithms with transition probabilities given by eq. 2.7, unlike e.g. Romeo and Sangiovanni-Vincentelli [ROM85], who distinguish between *probabilistic hill climbing* (general generation and acceptance matrices) and the simulated annealing algorithm in its original formulation.
As pointed out before, the control parameter c is decreased during the course of the algorithm. With respect to this decrement, the following two formulations of the algorithm can be distinguished:

- a *homogeneous algorithm*: the algorithm is described by a sequence of homogeneous Markov chains. Each Markov chain is generated at a fixed value of c and c is decreased in between subsequent Markov chains, and

- an *inhomogeneous algorithm*: the algorithm is described by a single inhomogeneous Markov chain. The value of c is decreased in between subsequent transitions.

The simulated annealing algorithm obtains a global minimum if, after a (possibly large) number of transitions, say K, the following relation holds:

$$Pr\{\mathbf{X}(K) \in \mathcal{R}_{opt}\} = 1, \tag{2.8}$$

where $\mathcal{R}_{opt}$ is the set of *globally minimal configurations*. In the next chapter, it is shown that for the homogeneous algorithm eq. 2.8 holds asymptotically (i.e. $\lim_{K\to\infty} Pr\{\mathbf{X}(K) \in \mathcal{R}_{opt}\} = 1$), if

1. each individual Markov chain is of infinite length;

2. certain conditions on the matrices $A(c_l)$ and $G(c_l)$ are satisfied;

3. $$\lim_{l\to\infty} c_l = 0, \tag{2.9}$$

where c_l is the value of the control parameter of the l-th Markov chain. For the inhomogeneous algorithm, eq. 2.8 holds asymptotically if

1. certain conditions on the matrices $A(c_k)$ and $G(c_k)$ are satisfied;

2. $$\lim_{k \to \infty} c_k = 0; \quad (2.10)$$

3. under certain additional conditions on the matrix $A(c_k)$, the rate of convergence of the sequence $\{c_k\}$ is not faster than $\mathcal{O}([\log k]^{-1})$.

These two convergence results are extensively discussed in chapter 3.

Chapter 3

Asymptotic convergence results

3.1 The homogeneous algorithm

3.1.1 Introduction

Essential to the convergence proof for the homogeneous algorithm is the fact that, under certain conditions, the *stationary distribution* of a homogeneous Markov chain exists. The stationary distribution is defined as the vector **q** whose i-th component is given by [FELL50]

$$q_i = \lim_{k \to \infty} Pr\{\mathbf{X}(k) = i \mid \mathbf{X}(0) = j\}, \tag{3.1}$$

for an arbitrary j.
From eq. 3.1 it follows that

$$q_i = \lim_{k \to \infty} Pr\{\mathbf{X}(k) = i\} = \lim_{k \to \infty} \mathbf{a}(0)^T P^k, \tag{3.2}$$

where $\mathbf{a}(0)$ denotes the *initial probability distribution*, i.e. $\mathbf{a}(0) = (a_i(0))$, $i \in \mathcal{R}$ satisfies

$$\forall i \in \mathcal{R} : a_i(0) \geq 0, \sum_{i \in \mathcal{R}} a_i(0) = 1. \tag{3.3}$$

Thus, the stationary distribution is the probability distribution of the configurations after an infinite number of transitions. Clearly, as for simulated annealing P depends on c, $\mathbf{q}$ depends on c, i.e. $\mathbf{q} = \mathbf{q}(c)$. The convergence proof is now based on the following arguments ([AAR85a], [LUN86], [ROM85], [ROSS86]). First, conditions on the matrices $A(c)$ and $G(c)$, defining the homogeneous Markov chain, are derived such that the stationary distribution $\mathbf{q}(c)$ exists. Next, these conditions are refined such that for decreasing c, $\mathbf{q}(c)$ converges to a uniform distribution on the set of globally minimal configurations, i.e.

$$\lim_{c \downarrow 0} \mathbf{q}(c) = \pi, \tag{3.4}$$

where the $| \mathcal{R} |$-vector π is defined by

$$\pi_i = \begin{cases} | \mathcal{R}_{opt} |^{-1} & \text{if } i \in \mathcal{R}_{opt} \\ 0 & \text{elsewhere,} \end{cases} \tag{3.5}$$

where $\mathcal{R}_{opt}$ is the set of globally minimal configurations.
Combining eqs. 3.2 and 3.4 then yields

$$\lim_{c \downarrow 0} (\lim_{k \to \infty} Pr\{\mathbf{X}(k) = i\}) = \pi_i, \tag{3.6}$$

where π_i is given by eq. 3.5. Thus, the proof of convergence is complete, since we now have

$$\lim_{c \downarrow 0} (\lim_{k \to \infty} Pr\{\mathbf{X}(k) \in \mathcal{R}_{opt}\}) = 1. \tag{3.7}$$

In the remainder of this section, the conditions on $G(c)$ and $A(c)$ to ensure the existence of $\mathbf{q}(c)$ are discussed in subsection 3.1.2 and they are further refined in subsection 3.1.3.

3.1.2 Existence of the stationary distribution

The following theorem establishes the existence of the stationary distribution.

Theorem 1 (Feller, [FELL50])
The stationary distribution $\mathbf{q}$ *of a finite homogeneous Markov chain*

exists if the Markov chain is irreducible *and* aperiodic. *Furthermore, the vector* $\mathbf{q}$ *is uniquely determined by the following equations:*

$$\forall i : q_i > 0, \sum_i q_i = 1, \tag{3.8}$$

$$\forall i : q_i = \sum_j q_j P_{ji}. \tag{3.9}$$

Note that $\mathbf{q}$ is the left *eigenvector* of the matrix P with *eigenvalue* 1. A Markov chain is

1. *irreducible*, if and only if for all pairs of configurations (i, j) there is a positive probability of reaching j from i in a finite number of transitions, i.e.

$$\forall i, j \; \exists n : 1 \leq n < \infty \; \wedge \; (P^n)_{ij} > 0; \tag{3.10}$$

2. *aperiodic*, if and only if for all configurations $i \in \mathcal{R}$, the greatest common divisor of all integers $n \geq 1$, such that

$$(P^n)_{ii} > 0 \tag{3.11}$$

is equal to 1.

In the case of simulated annealing, the matrix P is defined by eq. 2.7. Since we assume that $\forall i, j, c > 0 : A_{ij}(c) > 0$ (see subsection 3.1.3), it is sufficient for irreducibility to assume that the Markov chain induced by $G(c)$ is irreducible itself, i.e. [AAR85a], [LUN86], [ROM85], [ROSS86]

$$\forall i, j \in \mathcal{R} \; \exists p \geq 1 \; \exists l_0, l_1, \ldots, l_p \in \mathcal{R} \; (l_0 = i \wedge l_p = j) :$$

$$G_{l_k l_{k+1}}(c) > 0, \; k = 0, 1, \ldots, p-1. \tag{3.12}$$

To establish aperiodicity, we use the fact that an irreducible Markov chain is aperiodic if the following condition is satisfied [ROM85]:

$$\forall c > 0 \; \exists i_c \in \mathcal{R} : P_{i_c i_c}(c) > 0. \tag{3.13}$$

Thus, for aperiodicity it is sufficient to assume that [ROM85]

$$\forall c > 0 \; \exists i_c, j_c \in \mathcal{R} : A_{i_c j_c}(c) < 1 \wedge G_{i_c j_c} > 0, \tag{3.14}$$

as we now have, using the inequalities of eq. 3.14 and the fact that $\forall i, j : A_{ij} \leq 1$:

$$\sum_{l=1, l \neq i_c}^{|\mathcal{R}|} A_{i_c l}(c) G_{i_c l}(c) = \sum_{l=1, l \neq i_c, j_c}^{|\mathcal{R}|} A_{i_c l}(c) G_{i_c l}(c) + A_{i_c j_c}(c) G_{i_c j_c}(c) <$$

$$\sum_{l=1, l \neq i_c, j_c}^{|\mathcal{R}|} G_{i_c l}(c) + G_{i_c j_c}(c) = \sum_{l=1, l \neq i_c}^{|\mathcal{R}|} G_{i_c l}(c) \leq \sum_{l=1}^{|\mathcal{R}|} G_{i_c l}(c) = 1. \tag{3.15}$$

Thus

$$P_{i_c i_c} = 1 - \sum_{l=1, l \neq i_c}^{|\mathcal{R}|} A_{i_c l}(c) G_{i_c l}(c) > 0 \tag{3.16}$$

and thus, eq. 3.13 holds for $i = i_c$.
Note that in the initial formulation of the algorithm ([KIR82], [ČER85]), the acceptance probabilities are defined by

$$A_{ij}(c) = \min\{1, \exp(-(C(j) - C(i))/c)\} \tag{3.17}$$

and hence eq. 3.14 is always satisfied by setting, for all $c > 0$, $i_c \in \mathcal{R}_{opt}, j_c \notin \mathcal{R}_{opt}$.
Summarizing we have the following result. The homogeneous Markov chain with conditional probabilities given by eq. 2.7 has a stationary distribution if the matrices $A(c)$ and $G(c)$ satisfy eqs. 3.12 and 3.14, respectively.

3.1.3 Convergence of the stationary distribution

We now impose further conditions on the matrices $A(c)$ and $G(c)$ to ensure convergence of $\mathbf{q}(c)$ to the distribution π, as given by eq. 3.5. The most general and least restrictive set of such conditions is derived by Romeo and Sangiovanni-Vincentelli [ROM85]. These restrictions are least restrictive in the sense that all other conditions, as derived by Aarts and Van Laarhoven [AAR85a], Anily and Federgrun [ANI87a],

Lundy and Mees [LUN86] and Otten and Van Ginneken [OTT84], imply the conditions of Romeo *et al.*, but the reverse does not hold. The derivation by Romeo and Sangiovanni-Vincentelli [ROM85] is based on the fact that for an arbitrary configuration $i \in \mathcal{R}$ the corresponding component of the stationary distribution can be written as

$$q_i(c) = \frac{\psi(C(i), c)}{\sum_j \psi(C(j), c)}, \tag{3.18}$$

where $\psi(\gamma, c)$ is a two-argument function, provided $\psi(\gamma, c)$ satisfies the following two conditions:

1. $$\forall i \in \mathcal{R}, c > 0 : \psi(C(i), c) > 0 \tag{3.19}$$
2. a *global balance* requirement, viz. $\forall j \in \mathcal{R}$:
$$\sum_{i=1, i \neq j}^{|\mathcal{R}|} \psi(C(i), c) G_{ij}(c) A_{ij}(c) = \psi(C(j), c) \sum_{i=1, i \neq j}^{|\mathcal{R}|} G_{ji}(c) A_{ji}(c). \tag{3.20}$$

Note that $\mathbf{q}(c)$, as defined by eq. 3.18, is indeed the unique stationary distribution, since the $q_i(c)$'s satisfy eqs. 3.8 and 3.9 (the latter can be shown by using eq. 3.20).
To ensure $\lim_{c \downarrow 0} \mathbf{q}(c) = \pi$, the following conditions on the function $\psi(\gamma, c)$ are now sufficient [ROM85]:

1. $$\lim_{c \downarrow 0} \psi(\gamma, c) = \begin{cases} 0 & \text{if } \gamma > 0 \\ \infty & \text{if } \gamma < 0; \end{cases} \tag{3.21}$$
2. $$\frac{\psi(\gamma_1, c)}{\psi(\gamma_2, c)} = \psi(\gamma_1 - \gamma_2, c); \tag{3.22}$$
3. $$\forall c > 0 : \psi(0, c) = 1. \tag{3.23}$$

Summarizing, we have established that the conditions to be imposed on the matrices $A(c)$ and $G(c)$ to ensure asymptotic convergence to a global minimum are given by eqs. 3.12, 3.14 and 3.20, where in eq. 3.20, ψ is a positive two-argument function, satisfying eqs. 3.21-3.23.
The conditions on the $q_i(c)$'s, given by eqs. 3.18-3.23, are sufficient but not necessary to define the stationary distribution of the Markov chain, as given by eqs. 3.8 and 3.9. Furthermore, it is hard to establish

an explicit form for the stationary distribution using eqs. 3.18-3.23. For this reason a number of authors concentrate on a special choice for the two-argument function $\psi(\gamma, c)$, which results in more explicit conditions for the $q_i(c)$'s (from which an explicit form for $\mathbf{q}(c)$ is easier to derive) at the cost of a more restrictive set of conditions on the matrices $G(c)$ and $A(c)$. This choice, where $\psi(C(i) - C_{opt}, c)$ is taken as $A_{i_0 i}(c)$ (for an arbitrary configuration $i_0 \in \mathcal{R}_{opt}$) and $G(c)$ is not depending on c, is considered by Aarts and Van Laarhoven [AAR85a], Anily and Federgruen [ANI87a], Lundy and Mees [LUN86] and Otten and Van Ginneken [OTT84].

Theorem 2 (Folklore)
If the two-argument function $\psi(C(i) - C_{opt}, c)$ is taken as $A_{i_0 i}(c)$ (for an arbitrary configuration $i_0 \in \mathcal{R}_{opt}$) and if $G(c)$ is not depending on c, then the stationary distribution $\mathbf{q}(c)$ is given by

$$\forall i \in \mathcal{R} : q_i(c) = \frac{A_{i_0 i}(c)}{\sum_{j \in \mathcal{R}} A_{i_0 j}(c)}, \tag{3.24}$$

provided the matrices $A(c)$ and G satisfy the following conditions:

$$(a1) \qquad \forall i, j \in \mathcal{R} : G_{ji} = G_{ij}; \tag{3.25}$$

$$(a2) \qquad \forall i, j, k \in \mathcal{R} :$$

$$C(i) \leq C(j) \leq C(k) \Rightarrow A_{ik}(c) = A_{ij}(c) A_{jk}(c); \tag{3.26}$$

$$(a3) \qquad \forall i, j \in \mathcal{R} : C(i) \geq C(j) \Rightarrow A_{ij}(c) = 1; \tag{3.27}$$

$$(a4) \quad \forall i, j \in \mathcal{R}, c > 0 : C(i) < C(j) \Rightarrow 0 < A_{ij}(c) < 1, \tag{3.28}$$

The theorem is proved as follows. First, we remark that

$$\sum_j q_j(c) P_{ji}(c) = \sum_{j \neq i, C(j) \leq C(i)} \frac{1}{N} A_{i_0 j}(c) G_{ji} A_{ji}(c) +$$

$$\sum_{j \neq i, C(j) > C(i)} \frac{1}{N} A_{i_0 j}(c) G_{ji} A_{ji}(c) + q_i(c) P_{ii}(c) =$$

$$\sum_{j \neq i, C(j) \leq C(i)} \frac{1}{N} A_{i_0 i}(c) G_{ij} + \sum_{j \neq i, C(j) > C(i)} \frac{1}{N} A_{i_0 j}(c) G_{ij} + q_i(c) P_{ii}(c) =$$

$$q_i(c) \sum_{j \neq i, C(j) \leq C(i)} G_{ij} + \sum_{j \neq i, C(j) > C(i)} q_j(c) G_{ij} + q_i(c) P_{ii}(c) \tag{3.29}$$

and

$$q_i(c) P_{ii}(c) =$$

$$q_i(c) \left(1 - \sum_{j \neq i, C(j) \leq C(i)} G_{ij} A_{ij}(c) - \sum_{j \neq i, C(j) > C(i)} G_{ij} A_{ij}(c) \right) =$$

$$q_i(c) - q_i(c) \sum_{j \neq i, C(j) \leq C(i)} G_{ij} - \sum_{j \neq i, C(j) > C(i)} \frac{1}{N} A_{i_0 i}(c) G_{ij} A_{ij}(c) =$$

$$q_i(c) - q_i(c) \sum_{j \neq i, C(j) \leq C(i)} G_{ij} - \sum_{j \neq i, C(j) > C(i)} q_j(c) G_{ij}, \tag{3.30}$$

where N is the denominator in eq. 3.24. Combining eqs. 3.29 and 3.30 yields

$$\forall i \in \mathcal{R} : \sum_j q_j(c) P_{ji}(c) = q_i(c). \tag{3.31}$$

Furthermore, using eqs. 3.25-3.28, it is straightforward to show that $\mathbf{q}(c)$, as defined by eq. 3.24, satisfies eqs. 3.8 and 3.9.

□

Note that eqs. 3.25-3.28 imply eq. 3.20 (with $\psi(C(i)) = \psi(C_{opt}, c) \cdot A_{i_0 i}(c)$) but that the reverse is not true.
Furthermore, it is implicitly assumed that the acceptance probabilities depend only on the cost values of the configurations and not on the configurations themselves. Hence, $A_{i_0 i}(c)$ does not depend on the particular choice for i_0, since

$$\forall i_0 \in \mathcal{R}_{opt} : C(i_0) = C_{opt}. \tag{3.32}$$

Besides, to ensure that $\lim_{c \downarrow 0} \mathbf{q}(c) = \pi$, the following condition is sufficient:

$$(a5) \quad \forall i, j \in \mathcal{R} : C(i) < C(j) \Rightarrow \lim_{c \downarrow 0} A_{ij}(c) = 0, \tag{3.33}$$

since eqs. 3.27 and 3.33 guarantee that $\lim_{c \downarrow 0} \mathbf{q}(c) = \pi$.
Lundy and Mees [LUN86] show that condition $(a1)$ (eq. 3.25) can be replaced by

$$(a1)' \quad \forall i \in \mathcal{R} : G_{ij} = \begin{cases} |\mathcal{R}_i|^{-1} & \text{if } j \in \mathcal{R}_i \\ 0 & \text{elsewhere,} \end{cases} \tag{3.34}$$

where $\mathcal{R}_i$ is again the neighbourhood of configuration i, $\mathcal{R}_i = \{j \in \mathcal{R} \mid G_{ij} \neq 0\}$. In this case the stationary distribution is given by

$$q_i(c) = \frac{|\mathcal{R}_i| A_{i_0 i}(c)}{\sum_{j \in \mathcal{R}} |\mathcal{R}_j| A_{i_0 j}(c)}. \tag{3.35}$$

If G is symmetric **and** eq. 3.34 holds, then it is possible to show that $|\mathcal{R}_i|$ is necessarily independent of i [AAR86a]; G is then given by

$$G_{ij} = \begin{cases} R^{-1} & \text{if } j \in \mathcal{R}_i \\ 0 & \text{elsewhere,} \end{cases} \tag{3.36}$$

where $R = |\mathcal{R}_i|$, $\forall i \in \mathcal{R}$. In this case the stationary distribution is again given by eq. 3.24.
Another alternative to condition $(a1)$ is formulated by Anily and Federgruen [ANI87a]:

$$(a1)'' \quad \exists\, |\mathcal{R}| \times |\mathcal{R}| - \text{matrix } Q \text{ for which}$$

$$\forall i, j \in \mathcal{R} : Q_{ij} = Q_{ji}; \tag{3.37}$$

$$\forall i \in \mathcal{R}, j \in \mathcal{R}_i : G_{ij} = \frac{Q_{ij}}{\sum_{l \in \mathcal{R}} Q_{il}}, \tag{3.38}$$

in which case the stationary distribution is given by

$$q_i(c) = \frac{(\sum_l Q_{il}) A_{i_0 i}}{\sum_j (\sum_l Q_{jl}) A_{i_0 j}}. \tag{3.39}$$

In both cases ($(a1)'$ and $(a1)''$), eqs. 3.27 and 3.33 are still sufficient to ensure convergence to a global minimum.
We mention that in the initial formulation of the algorithm ([KIR82], [ČER85]), where $A(c)$ and G are given by eqs. 3.17 and 3.36, respectively, conditions $(a1)$-$(a5)$ are satisfied. Hence, the Markov chain associated with $A(c)$ and G is irreducible and aperiodic and the stationary distribution is given by:

$$q_i(c) = \frac{\exp(-(C(i) - C_{opt})/c)}{\sum_{j \in \mathcal{R}} \exp(-(C(j) - C_{opt})/c)}. \tag{3.40}$$

Note that conditions $(a1)$-$(a5)$ are **sufficient** but not **necessary** to ensure that the conditions of theorem 1 are satisfied. Thus, there may be acceptance and generation matrices **not** satisfying these conditions and still ensuring the existence of the stationary distribution. An example of such an acceptance matrix is the one used by Aarts and Korst [AAR86c] and by Ackley *et al.* [ACK85] (see also subsection 7.3.2, eq. 7.33), where $A(c)$ is given by

$$A_{ij}(c) = \left(1 + \exp\left(-\frac{C(j) - C(i)}{c}\right)\right)^{-1}. \tag{3.41}$$

This acceptance matrix does not satisfy conditions $(a2)$ and $(a3)$ (eqs. 3.26 and 3.27), but it can be shown to lead to a stationary distribution given by eq. 3.40, by showing that eq. 3.9 is satisfied, substituting the expressions for $A(c)$ and $\mathbf{q}(c)$ given by eqs. 3.41 and 3.40, respectively.

As an aside, we discuss some results and conditions derived and obtained by Rossier *et al.* [ROSS86]. They follow a slightly different approach to construct the Markov chain. Given a positive vector $\mathbf{q}(c)$, satisfying

$$\forall i, j \in \mathcal{R} : C(i) \leq C(j) \Rightarrow q_i(c) \geq q_j(c) \tag{3.42}$$

and

$$\forall c > 0 : \sum_{i \in \mathcal{R}} q_i(c) = 1, \tag{3.43}$$

a Markov chain is constructed, whose transition matrix is defined as

$$P_{ij}(c) = \begin{cases} G_{ij} \min\{1, \frac{q_j(c)}{q_i(c)}\} & \forall j \neq i \\ 1 - \sum_{l=1, l \neq i}^{|\mathcal{R}|} G_{il} \min\{1, \frac{q_l(c)}{q_i(c)}\} & j = i, \end{cases} \tag{3.44}$$

where G is a $|\mathcal{R}| \times |\mathcal{R}|$-matrix, whose corresponding Markov chain is irreducible.
Let $\Omega_i = \{j \in \mathcal{R} : C(j) = C(i)\}$, then the following conditions are necessary and sufficient to ensure that $\mathbf{q}(c)$ is the stationary distribution of the Markov chain induced by the transition matrix given by eq. 3.44 [ROSS86]:

1. $\forall i \in \mathcal{R} : \sum_{k \in \Omega_i} (G_{ik} - G_{ki}) = 0;$ (3.45)
2. $\forall i, j \in \mathcal{R} : C(i) \neq C(j) \Rightarrow G_{ij} = G_{ji}.$ (3.46)

Finally, to guarantee the convergence of $\mathbf{q}(c)$ to the uniform distribution on the set of optimal configurations, the following condition is sufficient [ROSS86]:

$$\forall i \in \mathcal{R} \setminus \mathcal{R}_{opt.} \exists j \in \mathcal{R} : C(j) < C(i) \wedge G_{ij} > 0. \tag{3.47}$$

The approach of Rossier *et al.* to construct a Markov chain with a given stationary distribution as described above is a well-known technique in the theory of Markov chains.
We end this section with three remarks:

1. Another sufficient, but not necessary condition to satisfy eq. 3.9 is given by the requirement that the system is *locally balanced*, i.e.

$$\forall i, j \in \mathcal{R} : q_i(c)P_{ij}(c) = q_j(c)P_{ji}(c). \tag{3.48}$$

   This equation is known in physics as the *detailed balance* equation [TOD83], [WOOD68]. Clearly, the global balance requirement, defined by eq. 3.20, is less restrictive than the requirement of local balance, defined by eq. 3.48.

2. Using eq. 3.48, Greene and Supowit [GREE86] offer the following motivation for the choice of the acceptance matrix, given by eq. 3.17, in the original formulation of the algorithm. Given any acceptance matrix $A'_{ij}(c)$ for which the detailed balance equation is satisfied with the stationary distribution given by eq. 3.40, it can be shown that

$$\forall i, j \in \mathcal{R} : A'_{ij}(c) \leq A_{ij}(c), \tag{3.49}$$

   i.e. choosing $A(c)$ according to eq. 3.17 leads to the highest rate of acceptance of transitions.

3. Both Aarts and Van Laarhoven [AAR85a] and Rossier *et al.* [ROSS86] give arguments for the two-argument function ψ in eq. 3.18 (and hence the acceptance probabilities in the special case where $\psi(C(i) - C_{opt}, c) = A_{i_0 i}(c)$) to be of exponential form. These arguments are based on continuity conditions on the cost function and functional equation 3.26.

3.2 The inhomogeneous algorithm

3.2.1 Introduction

In the previous section it is shown that, under certain conditions on the matrices $A(c)$ and $G(c)$, the simulated annealing algorithm converges to a global minimum with probability 1 if for each value c_l of the control parameter ($l = 0, 1, 2, \ldots$), the corresponding Markov chain is of infinite length and if the c_l eventually converge to 0 for $l \to \infty$, i.e. the validity of the following equation is shown:

$$\lim_{c \downarrow 0}(\lim_{k \to \infty} Pr\{X(k) = i\}) = \lim_{c \downarrow 0} q_i(c) = \begin{cases} |\mathcal{R}_{opt}|^{-1} & \text{if } i \in \mathcal{R}_{opt} \\ 0 & \text{elsewhere.} \end{cases} \quad (3.50)$$

One is also interested in the convergence of the algorithm when the limits in the left-hand side of eq. 3.50 are taken along a path in the (c, k)-plane, i.e. when the value of the control parameter is changed after **each** transition and, consequently, when c is c_k. Thus, an inhomogeneous Markov chain is obtained, whose transition matrix $P(k-1, k)$ $(k = 0, 1, 2, \ldots)$ is defined by

$$P_{ij}(k-1, k) = \begin{cases} G_{ij}(c_k)A_{ij}(c_k) & \forall j \neq i \\ 1 - \sum_{l=1, l \neq i}^{|\mathcal{R}|} G_{il}(c_k)A_{il}(c_k) & j = i. \end{cases} \quad (3.51)$$

In subsections 3.2.2 and 3.2.3, it will be shown that the conditions for convergence to global minima not only relate to the matrices $G(c_k)$ and $A(c_k)$ but also impose restrictions on the way the current value of the control parameter, c_k, is changed into the next one, c_{k+1}. Hereinafter, we assume that the sequence $\{c_k\}$, $k = 0, 1, 2, \ldots$, satisfies the following two conditions:

$$1. \quad \lim_{k \to \infty} c_k = 0; \quad (3.52)$$

$$2. \quad c_k \geq c_{k+1}, \; k = 0, 1, \ldots. \quad (3.53)$$

Thus, we do not exclude the possibility that c_k is kept constant during a number of transitions, in which case we again obtain a homogeneous Markov chain, but of finite length.
It will be shown that under certain conditions on the acceptance matrix the rate of convergence of the sequence $\{c_k\}$ cannot be faster

than $\frac{\Gamma}{\log k}$, for some constant Γ, giving a bound on the value of c_k for each k. Geman and Geman [GEM84] were the first to obtain an explicit expression for Γ. The bound was subsequently refined by Anily and Federgruen [ANI87a], [ANI87b] and Mitra *et al.* [MIT86]. All results are obtained in a similar way, using ergodicity theorems for inhomogeneous Markov chains, and are discussed in subsection 3.2.2. This subsection also contains sufficient conditions, derived by Gelfand and Mitter [GEL85], to ensure convergence of the algorithm to an **arbitrary** set of configurations (which can be taken as the set of global minima) and the subsection concludes with additional arguments, given by Gidas [GID85b] and Hajek [HAJ85], for the particular form of the aforementioned bound.
Necessary and sufficient conditions are derived by Hajek in [HAJ88] and Gidas in [GID85a], using results obtained for continuous-time inhomogeneous Markov chains (see subsection 3.2.3). These conditions lead to an even stronger bound on the current value of the control parameter, c_k, than the bound of Mitra *et al.* [MIT86].

3.2.2 Sufficient conditions for convergence

We need the following two definitions:

Definition 1 (Seneta, [SEN81])
An inhomogeneous Markov chain is weakly ergodic *if for all $m \geq 1, i, j, l \in \mathcal{R}$:*

$$\lim_{k\to\infty} (P_{il}(m,k) - P_{jl}(m,k)) = 0, \tag{3.54}$$

where the transition matrix $P(m,k)$ is defined by:

$$P_{il}(m,k) = Pr\{\mathbf{X}(k) = l \mid \mathbf{X}(m) = i\}. \tag{3.55}$$

Definition 2 (Seneta, [SEN81])
An inhomogeneous Markov chain is strongly ergodic *if there exists a vector π, satisfying*

$$\sum_{i=1}^{|\mathcal{R}|} \pi_i = 1, \quad \forall i : \pi_i \geq 0, \tag{3.56}$$

such that for all $m \geq 1, i, j \in \mathcal{R}$:

$$\lim_{k \to \infty} P_{ij}(m, k) = \pi_j. \tag{3.57}$$

Thus, weak ergodicity implies that eventually the dependence of $\mathbf{X}(k)$ with respect to $\mathbf{X}(0)$ vanishes (loss of memory), whereas strong ergodicity implies *convergence in distribution* of the $\mathbf{X}(k)$; if eq. 3.57 holds, we have:

$$\lim_{k \to \infty} Pr\{\mathbf{X}(k) = j\} = \pi_j. \tag{3.58}$$

For a homogeneous Markov chain, there is no distinction between weak and strong ergodicity [KOZ62].
The following two theorems provide conditions for weak and strong ergodicity of inhomogeneous Markov chains:

Theorem 3 (Seneta, [SEN81])
An inhomogeneous Markov chain is weakly ergodic if and only if there is a strictly increasing sequence of positive numbers $\{k_l\}, l = 0, 1, 2, \ldots$, *such that*

$$\sum_{l=0}^{\infty} (1 - \tau_1(P(k_l, k_{l+1})) = \infty, \tag{3.59}$$

where $\tau_1(P)$, *the* coefficient of ergodicity *of an* $n \times n$*-matrix* P, *is defined as*

$$\tau_1(P) = 1 - \min_{i,j} \sum_{l=1}^{n} \min(P_{il}, P_{jl}). \tag{3.60}$$

Theorem 4 (Isaacson and Madsen, [ISA76])
An inhomogeneous Markov chain is strongly ergodic if it is weakly ergodic and if for all k there exists a vector $\pi(k)$ *such that* $\pi(k)$ *is an* eigenvector *with* eigenvalue 1 *of* $P(k-1, k)$, $\sum_{i=1}^{|\mathcal{R}|} |\pi_i(k)| = 1$ *and*

$$\sum_{k=0}^{\infty} \sum_{i=1}^{|\mathcal{R}|} |\pi_i(k) - \pi_i(k+1)| < \infty. \tag{3.61}$$

Moreover, if $\pi = \lim_{k \to \infty} \pi(k)$, *then* π *is the vector in definition 2, i.e.* π *satisfies*

$$\lim_{k \to \infty} P_{ij}(m, k) = \pi_j. \tag{3.62}$$

Under the assumptions of subsection 3.1.2 on the matrices $A(c)$ and $G(c)$, there exists an eigenvector $\mathbf{q}(c_k)$ of $P(k-1,k)$, for each $k \geq 0$ (the stationary distribution of the **homogeneous** Markov chain with transition matrix given by eq. 3.51). Furthermore, under the additional assumptions of subsection 3.1.3, $\lim_{k\to\infty} \mathbf{q}(c_k) = \pi$, where the $|\mathcal{R}|$-vector π is given by eq. 3.5, provided $\lim_{k\to\infty} c_k = 0$. Using theorem 4 with $\pi(k) = \mathbf{q}(c_k)$, strong ergodicity can now be proved by showing that the following two conditions are satisfied:

1. the Markov chain is weakly ergodic;
2. the $\mathbf{q}(c_k)$, $k = 0,1,2,\ldots$, satisfy eq. 3.61.

Using eqs. 3.5 and 3.58, we then have

$$\lim_{k\to\infty} Pr\{\mathbf{X}(k) \in \mathcal{R}_{opt}\} = 1. \tag{3.63}$$

For simulated annealing in its original formulation, where $\mathbf{q}(c_k)$ is given by

$$q_i(c_k) = \frac{\exp(-(C(i) - C_{opt})/c_k)}{\sum_{j=1}^{|\mathcal{R}|} \exp(-(C(j) - C_{opt})/c_k)}, \tag{3.64}$$

the validity of eq. 3.61 is shown by Geman and Geman [GEM84] and by Mitra *et al.* [MIT86]. Furthermore, these authors, as well as Anily and Federgruen [ANI87a], [ANI87b], use theorem 3 to derive a sufficient condition on the sequence $\{c_k\}, k = 0,1,2,\ldots$, to ensure weak ergodicity. In their respective approaches, Geman and Geman as well as Mitra *et al.* confine themselves to the original formulation of the algorithm (acceptance and generation matrices given by eqs. 3.17 and 3.36, respectively), whereas Anily and Federgruen prove asymptotic convergence for the general case (generation and acceptance matrices satisfying the conditions of section 3.1) and apply this result to the case where the acceptance matrix is again given by eq. 3.17.

As mentioned before, Geman and Geman [GEM84] were the first to obtain such a condition on the sequence $\{c_k\}$. They show that if

$$\exists k_0 \geq 2 \; \forall k \geq k_0 : c_k \geq \frac{|\mathcal{R}|\, \Delta C_{max}}{\log k}, \tag{3.65}$$

where $\Delta C_{max} = \max\{C(i) \mid i \in \mathcal{R}\} - \min\{C(i) \mid i \in \mathcal{R}\}$, then eq. 3.59 is satisfied for some sequence $\{c_k\}$, $k = 0, 1, 2, \ldots$, and hence, weak ergodicity is obtained.

A similar, but sharper bound is obtained by Anily and Federgruen [ANI87a], [ANI87b] in the following way. Let n be an integer such that there is a globally minimal configuration that can be reached from any other configuration in no more than n transitions (since the Markov chain induced by G is irreducible, n exists and clearly $n <\mid \mathcal{R} \mid$). Furthermore, let

$$\Delta = \max_{i,j}\{C(j) - C(i) \mid i \in \mathcal{R}, j \in \mathcal{R}_i, C(j) > C(i)\} \tag{3.66}$$

and

$$\underline{A}(c) = \min_{i,j}\{A_{ij}(c) \mid i \in \mathcal{R}, j \in \mathcal{R}_i\}. \tag{3.67}$$

Anily and Federgruen prove that for general acceptance and generation matrices (satisfying the conditions of subsections 3.1.2 and 3.1.3) convergence is obtained if

$$\sum_{k=1}^{\infty} (\underline{A}(c_{kn}))^n = \infty. \tag{3.68}$$

Applying this result to the special case, where the acceptance matrix is given by eq. 3.17, yields that if

$$\forall k \geq 2 : c_k \geq \frac{n\Delta}{\log k}, \tag{3.69}$$

then eq. 3.59 is satisfied for $k = i \cdot n$, $i = 1, 2, \ldots$. In fact, it suffices to assume that

$$\forall i : c_{in} \geq \frac{n\Delta}{\log(in)}. \tag{3.70}$$

Clearly,

$$n \cdot \Delta < |\mathcal{R}| \cdot \Delta C_{max}, \tag{3.71}$$

which proves that the bound given by Anily and Federgruen is sharper than the one given by Geman and Geman.

Finally, the sharpest bound is obtained by Mitra *et al.* [MIT86]. Let

$$\mathcal{R}_{max} = \{i \in \mathcal{R} \mid \forall j \in \mathcal{R}_i : C(j) \leq C(i)\} \tag{3.72}$$

be the set of all *locally maximal* configurations and let

$$r = \min_{i \in \mathcal{R} \setminus \mathcal{R}_{max}} \max_{j \in \mathcal{R}} d(i,j), \tag{3.73}$$

where $d(i,j)$ is the minimal number of transitions to reach j from i. r is an integer such that there is at least one non-locally maximal configuration that can be reached from any other configuration in no more than r transitions, namely the one for which the minimum in eq. 3.73 is attained. Hence, $r \leq n$. Mitra *et al.* show that if

$$\forall k \geq 2 : c_k \geq \frac{r\Delta}{\log k}, \tag{3.74}$$

then eq. 3.59 is satisfied for $k = i \cdot r$, $i = 1, 2, \ldots$ In fact, it suffices to assume that

$$\forall i : c_{ir} \geq \frac{r\Delta}{\log(ir)}. \tag{3.75}$$

Thus far, sufficient conditions for the algorithm to converge to the set of globally minimal configurations were presented. In [GEL85], Gelfand and Mitter derive sufficient conditions for convergence to an **arbitrary** set of configurations I, which can, of course, be taken as the set of global minima. We need the following entities to discuss these conditions.
For any pair of configurations, i and j, we define $\mathcal{T}_{ij}^{(d)}$ as the set of all chains of transitions $i = l_0 \rightarrow l_1 \ldots l_d = j$ of length d, for which $P_{l_\alpha l_{\alpha+1}} > 0$ $(\alpha = 0, 1, \ldots, d-1)$. For $i, j \in \mathcal{R}, d \in \mathbf{N}$ and $\tau \in \mathcal{T}_{ij}^{(d)}$ let

$$\Gamma(\tau) = \sum_{\alpha=0}^{d-1} \max(0, C(l_{\alpha+1}) - C(l_\alpha)) \tag{3.76}$$

and

$$\Gamma_{ij}^{(d)} = \begin{cases} \min_{\tau \in \mathcal{T}_{ij}^{(d)}} \Gamma(\tau) & \text{if } \mathcal{T}_{ij}^{(d)} \neq \emptyset \\ \infty & \text{elsewhere} \end{cases} \tag{3.77}$$

and finally

$$\Gamma_{ij} = \min_{d \in \mathbf{N}} \Gamma_{ij}^{(d)}. \tag{3.78}$$

Let J be the set of configurations not belonging to I, i.e. $J = \mathcal{R} \setminus I$. If an index in eqs. 3.76-3.78 is replaced by a set, then an additional minimization is to be performed over the elements of the set, e.g.

$$\Gamma_{iJ}^{(d)} = \min_{j \in J} \Gamma_{ij}^{(d)}. \tag{3.79}$$

In [GEL85], conditions on the generation matrix $G(c)$ and the sequence $\{c_k\}, k = 0, 1, 2, \ldots$, are derived to ensure

$$\lim_{k \to \infty} Pr\{\mathbf{X}(k) \in I\} = 1. \tag{3.80}$$

These conditions are given in the following theorem, where it is assumed that the acceptance matrix $A(c)$ is given by eq. 3.17.

Theorem 5 (Gelfand and Mitter, [GEL85])
The following conditions are sufficient to ensure the validity of eq. 3.80:

1. $\exists d \in \mathbf{N}\ \forall j \in J : \Gamma_{jI}^{(d)} = \Gamma_{jI};$ (3.81)
2. $\max_{j \in J} \Gamma_{jI} < \infty$ (3.82)
   (every $j \in J$ can reach some $i \in I$);
3. $\forall j \in J : \Gamma_{jI} < \Gamma_{Ij};$ (3.83)
4. *for k large enough, $c_k \geq \frac{\Gamma^*}{\log k}$, with*
   $\Gamma^* = \max_{j \in J} \Gamma_{jI}.$ (3.84)

If I is such that it can be written as

$$I = \{i \in \mathcal{R} \mid C(i) \leq U\} \tag{3.85}$$

for some constant U, then condition (1) of theorem 5 is always satisfied [GEL85]. Moreover, if additionally, the matrix $G(c)$ is symmetric, then the third condition in theorem 5 is satisfied as well.
Finally, we remark that if U is replaced by C_{opt} in eq. 3.85, again a sufficient condition for convergence to the set of global minima is obtained by applying theorem 5:

$$c_k \geq \frac{\Gamma'}{\log k} \tag{3.86}$$

with

$$\Gamma' = \max_{j \notin \mathcal{R}_{opt}} \Gamma_{j\mathcal{R}_{opt}}. \tag{3.87}$$

In subsection 3.2.3 this bound is compared with a bound obtained by Hajek in [HAJ88].

To conclude this section, we discuss two results, presented by Gidas [GID85b] and Hajek [HAJ85], respectively, which give further support for the postulate that a rate of convergence of the sequence $\{c_k\}, k = 0, 1, 2, \ldots$, not faster than $\mathcal{O}([\log k]^{-1})$ is sufficient for convergence to the set of global minima.

1. In [GID85b], Gidas considers the *continuous-time analogue* of a *discrete-time* inhomogeneous Markov chain. Using conditions for strong ergodicity of such a Markov chain yields the requirement that

   $$\forall t > 0 : c(t) \geq \frac{\gamma_0}{\log t}, \tag{3.88}$$

   for some constant γ_0 (t denotes the time), suffices to ensure convergence to the set of global minima [GID85b].

2. Another argument for a condition of the form

   $$\forall k \geq 0 : c_k \geq \frac{\Gamma}{\log k}, \tag{3.89}$$

   for some constant Γ depending on the structure of the problem (as the constants in eqs. 3.65, 3.69, 3.74 and 3.86) is given by Hajek [HAJ85]. Hajek introduces the notion of a *cup* as the set of configurations that can be reached from a local minimum in a finite number of cost-increasing transitions (for a precise definition, the reader is referred to the next subsection) and considers again the continuous-time analogue. If $c(t)$ is given by

   $$c(t) = \frac{\Gamma}{\log t}, \tag{3.90}$$

   for some constant Γ, then it can be shown that the expected time to leave a cup $\mathcal{V}$ is finite if $\Gamma > d(\mathcal{V})$, where $d(\mathcal{V})$ is the depth of a cup $\mathcal{V}$, suitably defined. Moreover, there is strong evidence that for $\Gamma < d(\mathcal{V})$ there is a positive probability that the cup will never be left. As a matter of fact, it will be shown in the next subsection that $\Gamma \geq D$, where D is the largest depth of any cup, is both necessary and sufficient for convergence to global minima.

3.2.3 Necessary and sufficient conditions for convergence

Necessary and sufficient conditions are derived by Hajek [HAJ88] and Gidas [GID85a]. First, we discuss Hajek's results.
We need the following definition:

Definition 3 (Hajek, [HAJ88])
A configuration j is called reachable at height L *from a configuration i, if there is a sequence of configurations $i = l_0, l_1, \ldots, l_p = j$, such that*

$$G_{l_k l_{k+1}}(c) > 0, \; k = 0, 1, \ldots, p-1 \tag{3.91}$$

and

$$C(l_k) \leq L, \; k = 0, 1, \ldots, p. \tag{3.92}$$

A cup is now defined as a subset $\mathcal{V}$ of the set of configurations such that for some number E the following is true:

$$\forall i \in \mathcal{V} : \mathcal{V} = \{j \in \mathcal{R} \mid j \text{ is reachable from } i \text{ at height } E\}. \tag{3.93}$$

For a cup $\mathcal{V}$, we define $\underline{\mathcal{V}}$ and $\overline{\mathcal{V}}$ as

$$\underline{\mathcal{V}} = \min\{C(i) \mid i \in \mathcal{V}\}, \tag{3.94}$$

$$\overline{\mathcal{V}} = \min\{C(j) \mid j \notin \mathcal{V} \wedge \exists i \in \mathcal{V} : G_{ij} > 0\}. \tag{3.95}$$

The depth $d(\mathcal{V})$ of a cup $\mathcal{V}$ is now defined as $\overline{\mathcal{V}} - \underline{\mathcal{V}}$. Thus, a local minimum can be seen as a configuration i such that no configuration j with $C(j) < C(i)$ is reachable at height $C(i)$ from i. The depth of a local minimum i is taken to be the smallest number $d(i)$ such that there is a configuration j with $C(j) < C(i)$ reachable at height $C(i) + d(i)$ from i. If i is a global minimum, we set $d(i) = +\infty$.
Hajek's result can now be formulated as follows:

Theorem 6 (Hajek, [HAJ88])
Suppose that the one-step transition matrix is given by eq. 3.51, where $A(c_k)$ is given by eq. 3.17, and that the generation matrix is independent of c and satisfies the following two conditions:

1. *the Markov chain associated with G is irreducible (cf. eq. 3.12);*
2. *for any real number L and any two configurations i and j, i is reachable at height L from j if and only of j is reachable at height L from i.*

Assume, furthermore, that eqs. 3.52 and 3.53 hold. If D is the maximum of depths $d(i)$ of all configurations i that are local but not global minima, then

$$\lim_{k \to \infty} Pr\{\mathbf{X}(k) \in \mathcal{R}_{opt}\} = 1, \tag{3.96}$$

if and only if

$$\sum_{k=1}^{\infty} \exp\left(-\frac{D}{c_k}\right) = \infty \tag{3.97}$$

(note that the term $\exp(-\frac{D}{c_k})$ relates to the probability of accepting a transition which corresponds to an increase in the cost function of value D). The result is proved by considering the continuous-time analogue of the discrete-time inhomogeneous Markov chain defined by eq. 3.51.

If c_k is of the form

$$c_k = \frac{\Gamma}{\log k} \tag{3.98}$$

(cf. the conditions on the sequence $\{c_k\}$ in subsection 3.2.2), then Hajek's result clearly implies that eq. 3.96 holds if and only if $\Gamma \geq D$. According to Gelfand and Mitter [GEL85], the constant D in eq. 3.97 can be written as

$$D = \max_{j \notin \mathcal{R}_{opt}} \left(\min_{i \in \mathcal{R}_{opt}} D_{ji} \right), \tag{3.99}$$

where D_{ij} is defined analogously to Γ_{ij} (see eqs. 3.76-3.78):

$$D_{ij} = \min_{d \in \mathbf{N}} \left(\min_{\tau \in \mathcal{T}_{ij}^{(d)}} \left(\max_{\alpha=0,\ldots,d-1} (\max(0, C(l_{\alpha+1}) - C(i))) \right) \right). \tag{3.100}$$

Since, for any i, we have

$$\max_{\alpha=0,\dots,d-1}(\max(0, C(l_{\alpha+1}) - C(i))) \leq \sum_{\alpha=0}^{d-1} \max(0, C(l_{\alpha+1}) - C(l_\alpha)), \tag{3.101}$$

it follows that $D \leq \Gamma'$, where Γ' is the constant found by Gelfand and Mitter [GEL85] (eqs. 3.86 and 3.87). Thus, Hajek's result is stronger than the one of Gelfand and Mitter, but the last authors claim that it is easier to check whether their conditions (conditions (1)-(4) of theorem 5) are satisfied than it is for Hajek's condition (eq. 3.97). Since Hajek's condition is both necessary and sufficient, whereas the conditions of Geman and Geman [GEM84], Anily and Federgruen [ANI87a], [ANI87b] and Mitra *et al.* [MIT86] are sufficient only, we conclude that D is smaller than the constants in eqs. 3.65, 3.69 and 3.74, respectively.

Gidas [GID85a] proves convergence results for a general class of inhomogeneous Markov chains and next applies these results to the Markov chain associated with the simulated annealing algorithm. It is shown that under certain conditions relating to the matrix Π, whose entries are defined by

$$\forall i, j \in \mathcal{R} : \Pi_{ij} = \lim_{k\to\infty} P_{ij}(k-1, k), \tag{3.102}$$

a necessary and sufficient condition for the annealing algorithm to converge with probability 1 to a global minimum is given by:

$$\exists k_0 \geq 1\ \forall k \geq k_0 : c_k \geq \frac{\Gamma}{\log k}, \tag{3.103}$$

where Γ is given by

$$\Gamma = \min_{\gamma=1,2}\left(\min_{i\in\overline{S}(\gamma),\, j\in\overline{R}}\left(\min_{\Lambda_{ij}}\left(\sum_{\alpha=1}^{p}\max(0, C(l_\alpha) - C(l_{\alpha-1}))\right)\right)\right). \tag{3.104}$$

Here, $\overline{S}(1)$, $\overline{S}(2)$ and $\overline{R}$ are subsets of $\mathcal{R}$ corresponding to certain submatrices $S(1)$, $S(2)$ and R of P, respectively. Λ_{ij} denotes the set of all finite chains of transitions $i = l_0 \to l_1, \dots, l_{p-1} \to l_p = j$, for which $G_{l_\alpha l_{\alpha+1}} > 0$ ($\alpha = 0, 1, \dots, p-1$) (cf. the definition of $\mathcal{T}_{ij}$ in

subsection 3.2.2 and the bound given by eqs. 3.86 and 3.87).
However, according to Hajek [HAJ86], the condition given by eqs. 3.103 and 3.104 is not correct, due to a mistake in the proof of Gidas' result. Indeed, for the example given by Gidas on page 117 of his paper [GID85a], eq. 3.104 leads to $\Gamma = \min(U_2 - U_1, U_4 - U_3)$, whereas Hajek's constant D (see theorem 6) is given by $D = (U_4 - U_3)$ [HAJ86].

To conclude this section, we summarize the main results. The convergence of the inhomogeneous algorithm to global minima is discussed in terms of conditions on the sequence $\{c_k\}$ and the generation matrix $G(c_k)$ and in terms of an acceptance matrix, given by eq. 3.17. Given the following expression for c_k:

$$c_k = \frac{\Gamma}{\log k}, \tag{3.105}$$

it is shown that these conditions lead to bounds on the constant Γ of the form $\Gamma \geq d$, where d is some parameter depending on the structure of the optimization problem. A number of increasingly smaller values of d are presented, for which $\Gamma \geq d$ is sufficient for convergence to the set of global minima, as well as an expression for d to ensure convergence to an arbitrary set of configurations. Finally, an expression for d is discussed which leads to a necessary and sufficient condition of the form $\Gamma \geq d$.

Chapter 4

The relation with statistical physics

4.1 Introduction

As pointed out in the introduction (chapter 1) there exists a clear analogy between the *annealing of solids* and the problem of solving large combinatorial optimization problems. The physical annealing process can be successfully modelled by using computer simulation methods from *condensed matter physics* [BAR76], [WOOD68]. These methods in turn are based on the theory of statistical mechanics which can be viewed as the central discipline of condensed matter physics [BIN78], [TOD83]. Starting off with the first paper on simulated annealing by Kirkpatrick *et al.* [KIR82], a number of authors elaborate on the relation between combinatorial optimization and statistical mechanics, either because of a phenomenological interest in the analogy (Bernasconi [BER87], Bonomi and Lutton [BON84], Kirkpatrick *et al.* [KIR82], [KIR83], [KIR85] and Mézard and Parisi [MÉZ85], [MÉZ86]) or because of a possible framework to model the convergence and the corresponding control of the simulated annealing algorithm (Aarts and Van Laarhoven [AAR85a], [AAR88a], Grest *et al.* [GRES86], Kirkpatrick *et al.* [KIR82], [KIR83], Otten and Van Ginneken [OTT84], Randelman and Grest [RAN86] and White [WHI84]). In this chapter a number of quantities are discussed that are defined in the theory of statistical mechanics for the description of a physical many-particle system at equilibrium and it will be elucidated how these quantities are related to the problem of solving large combinatorial optimization problems. Moreover, using these quantities as a guideline, a number of characteristic features of the simulated annealing algorithm are discussed. The discussion presented in this chapter focusses on global aspects such as ***entropy***, ***ensemble averages*** (discrete and continuous), ***phase transitions*** and the relation between ***spin-glasses*** and combinatorial optimization problems. More specific studies based on relations with physics such as

- the *configuration-space analysis* based on *ultra metricity* (Kirkpatrick and Toulouse [KIR85]),
- the *replica* analysis (Mézard *et al.* [MÉZ84], Mézard and Parisi [MÉZ85], [MÉZ86]) and the *cavity* method (Mézard [MÉZ87]) and,
- the optimization via the *Langevin equation* (Gidas [GID85b])

are not discussed in great detail in this monograph and the interested reader is referred to the original papers on this work. Bounds' review article [BOU86] is another good starting point for those who are interested in the relation between spin-glasses and simulated annealing.

4.2 Equilibrium dynamics

Starting off with the fundamental assumption of statistical physics that the mechanics of a physical many-particle system is compatible with a statistical ensemble and admitting that the time average of a mechanical quantity of the system under macroscopic equilibrium is equal to the corresponding ensemble average (***ergodicity hypothesis***) a number of useful macroscopic quantities can be derived given the equilibrium distribution of the system. As first mentioned by Gibbs [GIB02], if the ensemble is stationary, its density is a function of the energy of the system. Moreover, applying the principle of equal probability [TOD83], it can be shown that at thermal equilibrium the probability that the system is in a macroscopic state i with energy E_i

is given by the *Gibbs* or *Boltzmann distribution* (see eq. 2.1)

$$Pr\{\mathbf{E} = E_i\} = \frac{1}{Z(T)} \exp(-E_i/k_B T), \tag{4.1}$$

where T is the temperature of the system, k_B the Boltzmann constant and $Z(T)$ the *partition function* defined as

$$Z(T) = \sum_i \exp(-E_i/k_B T), \tag{4.2}$$

where the summation runs over all possible macroscopic states. The relation between statistical physics and optimization of combinatorial problems can now be made more explicit: given a physical system in thermal equilibrium whose internal states are distributed according to the expression given by eq. 4.1 and a combinatorial optimization problem whose configurations are distributed according to the expression given by eq. 2.2, (which is identical to eq. 4.1), the ***equilibrium distribution***, a set of macroscopic quantities can be defined for the optimization problem in a similar way as for the physical system. Thus the following quantities can be defined:

1. the *expected cost* in equilibrium

$$\langle C(c)\rangle = \sum_{i\in\mathcal{R}} C(i)q_i(c), \tag{4.3}$$

where $\mathbf{q}(c)$ is given by eq. 2.2;

2. the *expected square cost* in equilibrium

$$\langle C^2(c)\rangle = \sum_{i\in\mathcal{R}} C^2(i)q_i(c); \tag{4.4}$$

3. the *variance* in the cost at equilibrium

$$\sigma^2(c) = \langle (C(c) - \langle C(c)\rangle)^2\rangle = \langle C^2(c)\rangle - \langle C(c)\rangle^2; \tag{4.5}$$

4. the *entropy* at equilibrium

$$S(c) = -\sum_{i\in\mathcal{R}} q_i(c)\ln q_i(c). \tag{4.6}$$

It can be straightforwardly shown that the following relations hold

$$\frac{\partial}{\partial c}\langle C(c)\rangle = \frac{\sigma^2(c)}{c^2} \tag{4.7}$$

and

$$\frac{\partial}{\partial c}\langle C(c)\rangle = c\frac{\partial}{\partial c}S(c). \tag{4.8}$$

The quantity $\frac{\partial}{\partial c}\langle C(c)\rangle$ is known in statistical physics as the *specific heat*. Moreover, defining the analogue of the partition function as (see also eq. 4.2)

$$Q(c) = \sum_{i \in \mathcal{R}} \exp\left(-\frac{C(i)}{c}\right), \tag{4.9}$$

it can be shown that

$$\langle C(c)\rangle = -\frac{\partial \ln Q(c)}{\partial (c^{-1})} \tag{4.10}$$

and

$$\langle F(c)\rangle = -c \ln Q(c) = \langle C(c)\rangle - c \cdot S(c), \tag{4.11}$$

where $F(c)$ is the equivalent of the *Helmholtz free energy* [TOD83].

Eqs. 4.3-4.11 are well known in statistical physics and they serve an important role in the analysis of the mechanics of large physical ensembles at equilibrium. Hereinafter, we elaborate in more detail on a number of these quantities and discuss some aspects of interest to the analysis of the simulated annealing algorithm.

Provided the equilibrium distributions corresponding to the simulated annealing algorithm are given by the $\mathbf{q}(c)$ of eq. 2.2 it can be shown that

$$S(\infty) = \lim_{c \to \infty} S(c) = \ln |\mathcal{R}| \tag{4.12}$$

and

$$S(0) = \lim_{c \downarrow 0} S(c) = \ln |\mathcal{R}_{opt}| . \tag{4.13}$$

Using eqs. 4.7 and 4.8 it then follows that during execution of the simulated annealing algorithm the entropy decreases monotonically, if the system reaches equilibrium at each value of the control parameter;

finally, the entropy reaches the value $\ln|\mathcal{R}_{opt}|$. In physics, if it is assumed that there exists only one *ground state*, eq. 4.13 reduces to

$$\lim_{c\downarrow 0} S(c) = \ln(1) = 0, \tag{4.14}$$

which is known as the *third law of thermodynamics* [TOD83].

The entropy can be interpreted as a natural measure of the order of a physical system: high entropy values correspond to chaos; low entropy values to order [KIR82], [TOD83]. A similar definition for the entropy as given in eq. 4.6 is known in information theory [SHA48], where it is viewed as a quantitative measure of the information contents of a system. In the case of simulated annealing, or optimization in general, the entropy can be interpreted as a quantitative measure of the degree of optimality. This is illustrated by figure 4.1. The figure shows four configurations in the evolution of the optimization process carried out by simulated annealing for a 100-city travelling salesman problem (TSP) (see also chapter 6) for which the cities are located on the vertices of a regular square grid [AAR85a]. The initial configuration (figure 4.1a) is given by a random sequence of the 100 cities, which is far from an optimal configuration. The configuration looks very chaotic and the corresponding value of the entropy is large. In the course of the optimization process (figures 4.1b and 4.1c) the observed configurations are closer to a minimum; the configurations become less chaotic and the corresponding value of the entropy decreases. Finally a minimum is obtained (figure 4.1d) with a highly regular pattern for which the entropy is minimal.
Using eqs. 4.7 and 4.8 the entropy can be expressed as

$$S(c) = S(c_1) - \int_c^{c_1} \frac{\sigma^2(c')}{c'^3} dc', \tag{4.15}$$

for an arbitrarily chosen $c_1 > 0$. Usually one resorts to an estimate of $S(c_1)$ using the *high temperature approximation*, i.e. for large values of c_1, $S(c_1)$ can be approximated by [AAR85a]

$$S(c_1) = \ln|\mathcal{R}| - \frac{\sigma(c_1)}{2c_1^2}. \tag{4.16}$$

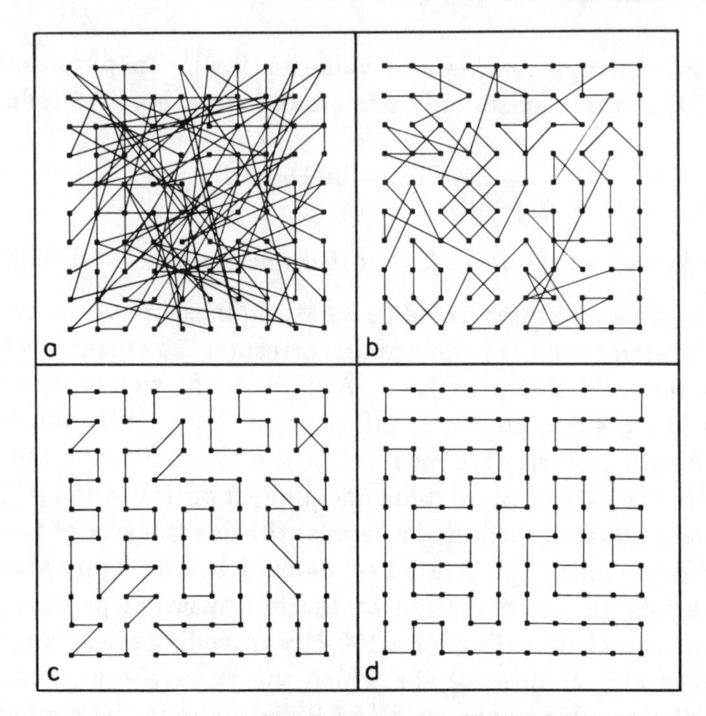

Figure 4.1: Configurations of a 100-city travelling salesman problem obtained after 0, 68, 92 and 123 steps of the algorithm. The initial tour looks very chaotic (a) (high entropy). Gradually, the tour becomes less chaotic, (b) and (c) (the entropy decreases). The final tour (d) shows a highly regular pattern (minimum cost and entropy) (reproduced from [AAR85b]).

The entropy can be helpful to estimate the minimum value of the cost function. In [ETT85], Ettelaie and Moore show how for a one-dimensional *Ising spin glass* the concept of *residual entropy* can be used to obtain useful information on the ground state of a spin glass.

4.3 Typical behaviour of the simulated annealing algorithm

In the case of the simulated annealing algorithm quantification of the macroscopic quantities given above can be done by approximating the expectations defined by eqs. 4.3-4.5. Using the central limit theorem and the law of large numbers it can be shown [AAR85a] that $\langle C(c)\rangle$ and $\langle C^2(c)\rangle$ can be approximated by the average of the cost function $\overline{C}(c)$ and the average of the square of the cost function $\overline{C^2}(c)$, respectively, sampled over the L_c configurations of the homogeneous Markov chain at the value c, i.e.

$$\langle C(c)\rangle \approx \overline{C}(c) = \frac{1}{L_c}\sum_{k=1}^{L_c} C(\mathbf{X}(k)) \tag{4.17}$$

and

$$\langle C^2(c)\rangle \approx \overline{C^2}(c) = \frac{1}{L_c}\sum_{k=1}^{L_c} C^2(\mathbf{X}(k)), \tag{4.18}$$

where $\mathbf{X}(k)$ denotes the configuration corresponding to the outcome of the k-th trial of the Markov chain.

Figure 4.2 shows typical examples of the dependence of $\overline{C}(c)$ (figure 4.2(a)) and $\sigma(c)$ (figure 4.2(b)), respectively, on the value of the control parameter c for the simulated annealing algorithm when applied to the travelling salesman problem used in figure 4.1. The data points are calculated according to eqs. 4.3 and 4.5 using the approximations given by eqs. 4.17 and 4.18. Clearly, one can observe that

$$\lim_{c\downarrow 0} \overline{C}(c) = C_{opt} \tag{4.19}$$

and

$$\lim_{c\downarrow 0} \sigma(c) = 0. \tag{4.20}$$

To model the typical behaviour of the simulated annealing algorithm as shown in figure 4.2 we briefly discuss an analytical approach to the calculation of the expected cost $\langle C(c)\rangle$ and the variance $\sigma^2(c)$. Let $\omega(C)dC$ be the *configuration density* defined as

$$\omega(C)dC = \frac{1}{|\mathcal{R}|}|\{i \in \mathcal{R} | C \leq C(i) < C + dC\}|. \tag{4.21}$$

Then in the case of the simulated annealing algorithm employing the acceptance probability of eq. 3.17, the *probability density* at a given

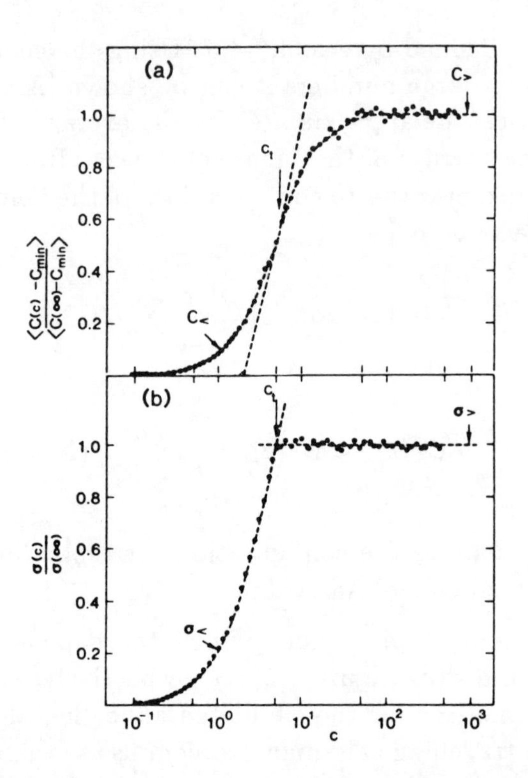

Figure 4.2: Normalized average value of the cost function (a) and the standard deviation of the cost function (b) as a function of the control parameter c for a 100-city TSP. The dashed lines are calculated according to eqs. 4.29 and 4.33 (a) and eqs. 4.30 and 4.34 (b) (reproduced from [AAR88a]).

value of c is given by

$$\Omega(C,c) = \frac{\omega(C)\exp(-(C - C_{opt})/c)}{\int_{-\infty}^{\infty}\omega(C')\exp((C' - C_{opt})/c)dC'}. \tag{4.22}$$

Clearly, $\Omega(C,c)$ is the equivalent of the stationary distribution $\mathbf{q}(c)$ given by eq. 2.2. The expected cost $\langle C(c)\rangle$ and the variance $\sigma^2(c)$ can now be written as

$$\langle C(c)\rangle = \int_{-\infty}^{\infty} C'\Omega(C',c)dC' \tag{4.23}$$

and

$$\sigma^2(c) = \int_{-\infty}^{\infty} (C' - \langle C(c)\rangle)^2 \Omega(C', c) dC'. \tag{4.24}$$

Given an analytical expression for the configuration density $\omega(C)$ it is possible to evaluate the integrals of eqs. 4.22-4.24. To estimate $\omega(C)$ for a given combinatorial optimization problem is in most cases very hard. Indeed $\omega(C)$ may vary drastically for different problems, especially for C-values close to C_{opt}. Analysis of the configuration space by using the concept of *ultra metricity* can be of use for some problems [KIR85], [SOL85]. However, some general statements can be made about the typical (i.e. average) behaviour of the configuration density. Numerical experiments carried out for large (pseudo-) random combinatorial problems support evidence for a typical behaviour that can be formulated by the following two postulates (see also [AAR88a], [HAJ85]):

Postulate 1 (Aarts *et al.* [AAR88a])
Let $\langle C(\infty)\rangle$ and $\sigma(\infty)$ be defined by

$$\langle C(\infty)\rangle = \lim_{c\to\infty} \langle C(c)\rangle \equiv \frac{1}{|\mathcal{R}|} \sum_{i\in\mathcal{R}} C(i) \tag{4.25}$$

and

$$\sigma^2(\infty) = \lim_{c\to\infty} \langle (C(c) - \langle C(\infty)\rangle)^2 \rangle \equiv \frac{1}{|\mathcal{R}|} \sum_{i\in\mathcal{R}} (C(i) - \langle C(\infty)\rangle)^2. \tag{4.26}$$

Then in the region of a few standard deviations $\sigma(\infty)$ around $\langle C(\infty)\rangle$ the configuration density is approximately Gaussian:

$$\omega(C) \propto \exp\left(-\frac{(C - \langle C(\infty)\rangle)^2}{2\sigma^2(\infty)}\right). \tag{4.27}$$

Consequently, in this region one obtains

$$\Omega(C, c) \propto \exp\left(\frac{-(C - (\langle C(\infty)\rangle - \frac{\sigma^2(\infty)}{c})^2}{2\sigma^2(\infty)}\right), \tag{4.28}$$

$$\langle C(c)\rangle \approx \langle C(\infty)\rangle - \frac{\sigma^2(\infty)}{c} \tag{4.29}$$

and

$$\sigma(c) \approx \sigma(\infty). \tag{4.30}$$

Postulate 2 (Aarts *et al.* [AAR88a])
In the region close to the minimum value of the cost function the configuration density can be approximated by

$$\omega(C) \propto \exp((C - C_{opt})\gamma), \tag{4.31}$$

for some constant $\gamma > 0$. *If* $\gamma < \frac{1}{c}$ *one obtains*

$$\Omega(C,c) \propto \left(\frac{1-\gamma c}{c}\right) \exp\left\{-(C - C_{opt})\left(\frac{1-\gamma c}{c}\right)\right\}, \tag{4.32}$$

$$\langle C(c) - C_{opt}\rangle \propto \frac{c}{1-\gamma c} \approx c + \gamma c^2 + \gamma^2 c^3 + \cdots \tag{4.33}$$

and

$$\sigma(c) \propto \frac{c}{1-\gamma c} \approx c + \gamma c^2 + \gamma^2 c^3 + \cdots. \tag{4.34}$$

From these postulates it follows that at high values of c the expected cost $\langle C(c)\rangle$ is proportional to $\frac{1}{c}$ and the standard deviation $\sigma(c)$ is constant ($C_>$ and $\sigma_>$ in figure 4.2), whereas at small values of c both $\langle C(c)\rangle$ and $\sigma(c)$ are proportional to c ($C_<$ and $\sigma_<$ in figure 4.2).

Clearly the meaning of "typical" is very vague. A possible way to give a meaning to the notion is to investigate large randomly generated problems. In section 6.3, for example, some results are discussed of a study carried out by Bonomi and Lutton on the typical behaviour for the travelling salesman problem and the quadratic assignment problem [BON84], [BON86].

4.4 Phase transitions

An interesting physical aspect in the study of the annealing algorithm is given by the phenomenon of *phase transitions* [MORI87]. Investigation of the analogy of phase transitions in a combinatorial optimization problem can be of help to study the convergence of the algorithm. A possible way to investigate a phase transition is by studying the specific heat of the system [WOOD68]. In figure 4.3 the specific heat is given for a *placement problem* (see subsection 7.2.2) calculated from eqs. 4.7 and 4.8 and using the approximations given by eqs. 4.17 and

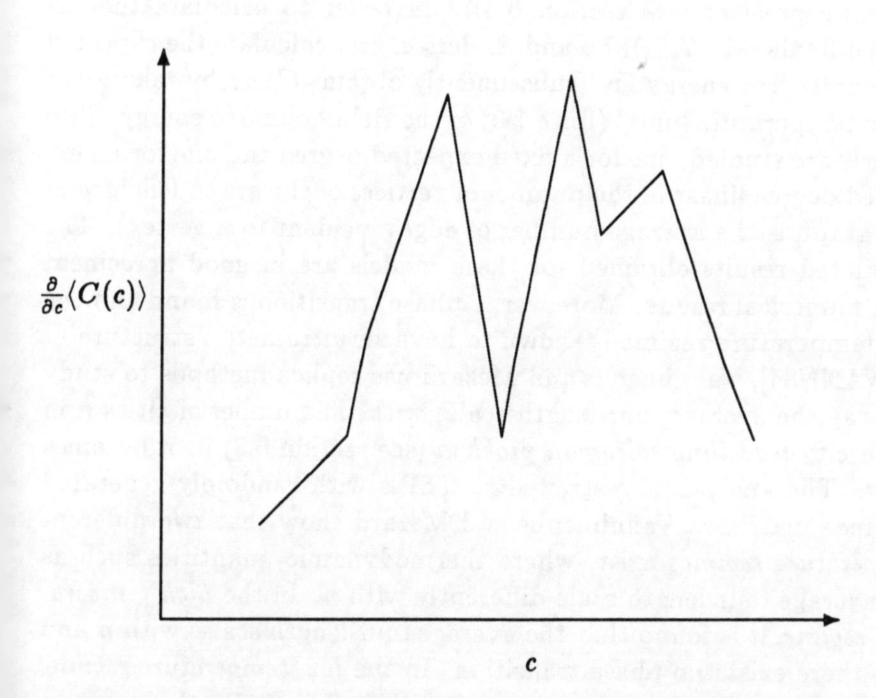

Figure 4.3: Specific heat, $\frac{\partial}{\partial c}\langle C(c)\rangle$, as a function of the control parameter for a placement problem (reproduced by permission from [KIR82]).

4.18 [KIR82]. This figure was obtained by Kirkpatrick *et al.* and they remark that, just as a maximum in the specific heat of a fluid indicates the onset of freezing or the formation of clusters, specific heat maxima are found at two values of the control parameter, each indicating a different type of ordering in the problem. This sort of measurement can be useful as a means of determining the ranges of c where a slower decrement in c will be helpful [KIR82].

Replica methods [MÉZ84] are successfully applied by a number of authors to study the general properties of combinatorial optimization problems, including the expected minimum cost and the existence and the nature of phase transitions. Within this scope, we briefly discuss two interesting papers.

In [FU86], Fu and Anderson apply replica analysis to the *graph partitioning problem* (see section 6.4). In order to calculate the expected final cost $\langle C_{min} \rangle$, Fu and Anderson first calculate the expected Helmholtz free energy and subsequently obtain $\langle C_{min} \rangle$ by taking the "zero temperature limit" ($\lim c \downarrow 0$) of the Helmholtz free energy. Two models are studied, one for a fixed expected degree and one for an expected degree linear in the number of vertices of the graph (the *degree* of a graph is the average number of edges incident to a vertex). The calculated results obtained for these models are in good agreement with numerical results. Moreover, a phase transition is found and the low temperature regime is shown to have an ultrametric structure.
In [VANN84], Vannimenus and Mézard use replica methods to study the way the average tour length scales with the number of cities n in the n-city *travelling salesman problem* (see section 6.3) as n becomes large. The analysis is restricted to TSP's with randomly generated distance matrices. Vannimenus and Mézard show that two different *temperature regimes* exist, where thermodynamic quantities such as the average tour length scale differently with n. In the *high temperature regime*, it is found that the average tour length scales with n and that there exists no phase transition. In the *low temperature regime*, a lower bound on the average tour length is found. Furthermore, in the special case where d-dimensional random point TSP's are considered, the average final tour length scales with $n^{1-\frac{1}{d}}$ for $d \to \infty$ (cf. the bound found by Beardwood *et al.* [BEA59], see eq. 6.11). Moreover, Vannimenus and Mézard find that for this class of problem instances, phase transitions exist in the low temperature regime.

4.5 Relation with spin glasses

As mentioned in section 2.1 there exists a relation between the energy of a physical system (represented by a *Hamiltonian*) and the cost function of a combinatorial optimization problem. It is pointed out by a number of authors (Fu and Anderson [FU86], Carnevalli *et al.* [CARN85], Kirkpatrick *et al.* [KIR82], Vecchi and Kirkpatrick [VEC83]) that there exists a striking analogy between the Hamiltonian of an Ising spin glass system and the cost function of several combinatorial optimization problems. As an illustrative example we discuss the cost function associated with the graph partitioning problem (see also section 6.4). Similar types of cost functions are also derived for other combinatorial problems, for example *routing* [VEC83] and *image processing* [CARN85].

In the graph partitioning problem the objective is to partition a graph into two subgraphs such that the difference in the number of vertices of the two subgraphs as well as the number of edges connecting the two subgraphs is as small as possible. Let a_{ij} denote the number of edges between two vertices i and j and μ_i a two-valued variable ($\mu_i = \pm 1$) indicating to which of the two subgraphs vertex i belongs. It is then easy to show that the number of edges running between the two subgraphs equals

$$\sum_{i>j} \frac{a_{ij}}{4}(\mu_i - \mu_j)^2 \tag{4.35}$$

and that the difference between the number of vertices of the two subgraphs equals

$$\sum_i \mu_i. \tag{4.36}$$

Squaring the second term and introducing a weighting factor λ the following cost function can be constructed (the constant terms are omitted)

$$C = \sum_{i>j} (\lambda - \frac{a_{ij}}{2}) \mu_i \mu_j. \tag{4.37}$$

This cost function has the same form as the Hamiltonian of a random magnet or spin glass, given by

$$H = \sum_{i>j} (J_0 - J_{ij}) s_i s_j, \tag{4.38}$$

where a spin s_i has two allowed orientations, i.e. "up" and "down", corresponding to $\mu_i = 1$ and $\mu_i = -1$, respectively. For such a system the Hamiltonian is composed of two terms, i.e. a short range attractive (ferromagnetic) component with coupling constants J_{ij} and a long range repulsive (anti-ferromagnetic) component with coupling constant J_0. These two terms are conflicting since they give rise to different ground states; there is no system configuration that can simultaneously satisfy these two ground states. This type of system is sometimes called *frustrated* [TOU77]. For the spin glass this implies that there are many degenerate ground states of about equal energy and that there is no obvious symmetry. Spin glasses provide a suggestive model for complex combinatorial optimization problems with a similar type of cost function. Kirkpatrick *et al.* [KIR82] formulate the following implications for the optimization by simulated annealing of problems with cost functions related to spin-glass Hamiltonians:

- Even in the presence of frustration, significant improvement can be obtained over iterative improvement starting from a randomly chosen configuration.

- There are many near-optimal solutions, so the probability that the simulated annealing algorithm will find one is large.

- All near-optimal solutions are very close to optimal solutions, so the search for a global minimum is not fruitful.

During the years extensive studies of spin glasses have been carried out emphasizing the aspect of frustration [MORI87]. Advances have directed attention to the distribution of local minima in the landscape of the configuration space (*configuration space analysis*) and the cooling rate dependence. With respect to the first item the work on ultrametricity is of special interest (see e.g. Kirkpatrick and Toulouse [KIR85], Parisi [PAR83], and Mézard *et al.* [MÉZ84]), e.g. evidence is presented for a hierarchical ultrametric structure of the configuration space of some combinatorial problems (TSP and graph partitioning). With respect to the second item, Grest *et al.* [GRES86] as well as Randelman and Grest [RAN86] conclude that the ground state energy obtained for certain NP-complete spin-glass models depends logarithmically on the cooling rate; a result which is theoretically confirmed by the asymptotic convergence properties presented in chapter 3. Similar results are obtained by Ettelaie and Moore [ETT85] from their residual entropy analysis.

Finally, we mention the connection between *neural-network models* and spin-glass models as an interesting aspect in the relation between combinatorial optimization and physics. Hopfield's model of content-addressable memories [HOP82] as well as the model of *Boltzmann machines* introduced by Hinton *et al.* [HIN83], [HIN84] (see also subsection 7.3.2) are essentially related to Ising spin models; the spins representing *neurons* and the interactions *synaptic* (connection) strengths. The learning algorithms associated with neural network models are based on the optimization of a cost function similar to a spin glass Hamiltonian. A physiologically appealing aspect in neural networks is the hierarchical structuring of knowledge representation since this provides a resourceful classification of memory and a more natural approach to the modelling of learning. Here again results obtained from hierarchical ultrametricity structuring of the configuration space corresponding to Ising-spin-like systems might prove useful. Moreover, recent work focusses on hardware implementations of combinatorial optimization problems [HOP85], [AAR87c], i.e. massively parallel decision networks are designed that embody the optimization problem such that after "optimization" (self-organization of the network) the system settles in a stable configuration which corresponds to a near-optimal solution of the optimization problem. Such a decision network can be viewed as a hardware implementation of an Ising-spin system. Thus it is tempting to explore ways to realize such a system based on the knowledge obtained from the study of Ising-spin models.

Chapter 5

Towards implementing the algorithm

5.1 Introduction

As is shown in chapter 3, the simulated annealing algorithm in its original formulation converges with probability 1 to a globally minimal configuration in either one of the following two cases:

1. for each value of the control parameter, c_k, an infinite number of transitions is generated and $\lim_{k \to \infty} c_k = 0$ (the homogeneous algorithm);

2. for each value c_k one transition is generated and c_k goes to zero not faster than $\mathcal{O}([\log k]^{-1})$ (the inhomogeneous algorithm).

In any implementation of the algorithm, asymptotic convergence can only be approximated. Thus, though the algorithm is asymptotically an optimization algorithm, any implementation results in an approximation algorithm. The number of transitions for each value c_k, for example, must be finite and $\lim_{k \to \infty} c_k = 0$ can only be approximated in a finite number of values for c_k. Due to these approximations, the algorithm is no longer guaranteed to find a global minimum with probability 1.
We remark that the convergence results of chapter 3 are not of great help when trying to approximate asymptotic convergence:

1. using results from Isaacson and Madsen ([ISA74]) and Seneta ([SEN81]), Aarts and Van Laarhoven [AAR85a] show that a sufficient condition on the *Markov chain length* L to ensure

$$\| \mathbf{a}(L, c) - \mathbf{q}(c) \| < \epsilon, \tag{5.1}$$

for an arbitrary small value of ϵ, where $\mathbf{a}(L, c)$ and $\mathbf{q}(c)$ are the probability distribution after L transitions and the stationary distribution of the homogeneous Markov chain at a value c of the control parameter, respectively, is given by

$$L > K(|\mathcal{R}|^2 + 3 \cdot |\mathcal{R}| + 3), \tag{5.2}$$

where K is proportional to $-\log \epsilon$. Usually, $|\mathcal{R}|$ is exponential in the size of the problem and eq. 5.2 then implies that the length of the Markov chains should at least be exponential in the problem size, which is, in the light of the discussion in chapter 1, highly undesirable. In section 6.2, a similar result is obtained for the inhomogeneous algorithm.

2. using the conditions of subsections 3.2.2 and 3.2.3 as a guideline for a rule to change the current value of the control parameter, c_k, into the next one, c_{k+1} (approximation of $\lim_{k \to \infty} c_k$) leads to a rule of the form

$$c_{k+1} = \frac{\Gamma}{\log\left(\exp\left(\frac{\Gamma}{c_k}\right) + 1\right)}, \tag{5.3}$$

which requires knowledge about the value of the constant Γ. Usually, however, it is extremely difficult to determine such a value (for small problems it can sometimes be determined by exhaustive enumeration). One resorts to conservative estimates, like $\Gamma = \Delta C_{max}$ (see eq. 3.66), which leads, however, to unnecessarily slow convergence of the algorithm (see e.g. the remarks concerning this aspect by Geman and Geman [GEM84] and by Lundy and Mees [LUN86]).

Of course, it is always possible to combine the homogeneous and inhomogeneous algorithms, by keeping the value of the control parameter constant during a number of transitions and taking care that the decrement in c_k is still not faster than $\mathcal{O}([\log k]^{-1})$. Such an approach is proposed by Rossier *et al.* ([ROSS86]) and leads to a decrement rule for c_k of the following type:

$$c_k = \gamma_i, \quad k_i \leq k < k_{i+1}, \tag{5.4}$$

$$\gamma_i = \frac{c}{\log k_i}, \quad i = 0, 1, 2, \ldots, \tag{5.5}$$

for some constant c and some sequence of positive numbers $\{k_i\}$, $i = 0, 1, 2, \ldots$. Note that eqs. 5.4 and 5.5 lead to a sequence of homogeneous Markov chains, the i-th chain consisting of k_i transitions.

Usually one resorts to an implementation of the simulated annealing algorithm in which a sequence of homogeneous Markov chains of finite length is generated at decreasing values of the control parameter. The following parameters should then be specified:

1. *initial value* of the control parameter, c_0;
2. *final value* of the control parameter, c_f (stop criterion);
3. *length* of Markov chains;
4. a rule for changing the current value of the control parameter, c_k, into the next one, c_{k+1}.

A choice for these parameters is referred to as a *cooling schedule*. In this chapter, several approaches to the problem of determining a cooling schedule are discussed. The discussion is confined to acceptance and generation matrices as given by eqs. 3.17 and 3.36, i.e. only the simulated annealing algorithm in its original formulation is considered. One of the few papers in which alternative acceptance matrices are considered is presented by Romeo and Sangiovanni-Vincentelli [ROM85], where it is concluded that the use of different acceptance matrices does not alter significantly the quality of solutions found by the algorithm.
Central in the construction of many cooling schedules is the concept of *quasi-equilibrium*: if L_k is the length of the k-th Markov chain, then the annealing algorithm is said to be in quasi-equilibrium at

c_k, the k-th value of the control parameter, if "**a**(L_k, c_k) **is close to q**(c_k)". The precise statement of this proximity is one of the major points differentiating one cooling schedule from the other. The actual construction of a cooling schedule is usually based on the following arguments.

- For $c_k \to \infty$, the stationary distribution is given by the uniform distribution on the set of configurations $\mathcal{R}$, which follows directly from eq. 3.64. Initially, quasi-equilibrium can therefore be achieved by choosing the initial value of c, c_0, such that virtually all transitions are accepted. For in that case, all configurations occur with equal probability, which corresponds to the aforementioned uniform distribution and thus to $\mathbf{q}(\infty)$.

- A *stop criterion* is usually based on the argument that execution of the algorithm can be terminated if the improvement in cost, to be expected in the case of continuing execution of the algorithm, is small.

- The length L_k of the k-th Markov chain and the transformation rule for changing c_k into c_{k+1} are strongly related through the concept of quasi-equilibrium. L_k is determined by specifying more precisely the meaning of "**a**(L_k, c_k) **is close to q**(c_k)". Concerning the transformation rule (usually a *decrement rule*), it is intuitively clear that large decrements in c_k will make it necessary to attempt more transitions at the new value of the control parameter, c_{k+1}, to restore quasi-equilibrium at c_{k+1}. For, given quasi-equilibrium at c_k, the larger the decrement in c_k, the larger the difference between $\mathbf{q}(c_k)$ and $\mathbf{q}(c_{k+1})$ and the longer it takes to establish quasi-equilibrium at c_{k+1} (this intuitive argument is quantified in section 6.4). Thus, there is a trade-off between fast decrement of c_k and small values for L_k. Usually, one opts for small decrements in c_k (to avoid extremely long chains), but alternatively, one could use large values for L_k in order to be able to make large decrements in c_k.

The search for adequate cooling schedules has been addressed in many papers during the last few years and several approaches are discussed

in the next two sections. Section 5.2 is devoted to schedules similar to the original schedule proposed by Kirkpatrick *et al.* in [KIR82]. These conceptually simple schedules are all based on empirical rules rather than on theoretically based choices. More elaborate and theoretically based schedules are discussed in section 5.3.
The chapter is concluded with a section in which two specific approaches to implementing the algorithm are discussed. These approaches both relate to improvements of the generation mechanism for transitions, resulting in faster execution of the algorithm.

5.2 Conceptually simple cooling schedules

In this section the cooling schedule as proposed by Kirkpatrick *et al.* [KIR82] and schedules of a similar nature are discussed. This discussion is gathered around the four parameters given in the previous section.

- initial value of the control parameter

The initial value of c is determined in such a way that virtually all transitions are accepted, i.e. c_0 is such that $\exp(-\Delta C_{ij}/c_0) \simeq 1$ for almost all i and j. Kirkpatrick *et al.* propose the following empirical rule: choose a large value for c_0 and perform a number of transitions. If the *acceptance ratio* χ, defined as the number of accepted transitions divided by the number of proposed transitions, is less than a given value χ_0 (in [KIR82] $\chi_0 = 0.8$), double the current value of c_0. Continue this procedure until the observed acceptance ratio exceeds χ_0.
This rule is further refined by a number of authors; Johnson *et al.* [JOHN87] determine c_0 by calculating the average increase in cost, $\overline{\Delta C}^{(+)}$, for a number of random transitions and solve c_0 from

$$\chi_0 = \exp(-\overline{\Delta C}^{(+)}/c_0). \tag{5.6}$$

Eq. 5.6 leads to the following choice for c_0

$$c_0 = \frac{\overline{\Delta C}^{(+)}}{\ln({\chi_0}^{-1})}. \tag{5.7}$$

A similar formula is proposed by Leong *et al.* [LEO85a], [LEO85b], Skiscim and Golden [SKI83], [GOLDE86], Morgenstern and Shapiro [MORC86], Aarts and Van Laarhoven [AAR85a], Lundy and Mees [LUN86] and Otten and Van Ginneken [OTT84], the latter three proposals belonging to the class of more elaborate schedules (to be discussed in section 5.3).

- final value of the control parameter

A stop criterion, determining the final value of the control parameter, is either determined by fixing the number of values c_k, for which the algorithm is to be executed (ranging from six, used by Nahar *et al.* [NAH85] to fifty, used by Bonomi and Lutton [BON84]) or by terminating execution of the algorithm if the last configurations of consecutive Markov chains are identical for a number of chains (Kirkpatrick *et al.* [KIR82], Morgenstern and Shapiro [MORC86], Sechen and Sangiovanni-Vincentelli [SEC85]). The latter criterion is refined by Johnson *et al.* [JOHN87]: they also require that the acceptance ratio is smaller than a given value χ_f.

- length of Markov chains

The simplest choice for L_k, the length of the k-th Markov chain, is a value depending (polynomially) on the size of the problem. Thus, L_k is independent of k. Such a choice is made by Bonomi and Lutton [BON84], [BON86], [LUT86], Burkard and Rendl [BUR84] and Sechen and Sangiovanni-Vincentelli [SEC85].

More elaborate proposals for the length of Markov chains are based on the intuitive argument that for each value c_k of the control parameter a minimum amount of transitions should be accepted, i.e. L_k is determined such that the number of accepted transitions is at least η_{min} (η_{min} some fixed number). However, as c_k approaches 0, transitions are accepted with decreasing probability and thus one eventually obtains $L_k \to \infty$ for $c_k \downarrow 0$. Consequently, L_k is ceiled by some constant L (usually chosen polynomial in the problem size) to avoid extremely

long Markov chains for low values of c_k.
Rules of this type are proposed by Kirkpatrick *et al.* [KIR82] (with $\overline{L} = n$, the number of variables of the problem to solve), Johnson et al. [JOHN87] (where $\overline{L} = m \cdot R$, a multiple of the size of the neighbourhoods), Leong *et al.* [LEO85a], [LEO85b] and Morgenstern and Shapiro [MORC86].
Nahar *et al.* [NAH85] determine L_k such that the number of rejected transitions is at least a fixed amount. Note that this approach leads to a gradually decreasing length of Markov chains during execution of the algorithm, which is the opposite of the previously described approach and in conflict with the intuitive arguments on which the previously described approach is based.
Intuitively closer to the concept of quasi-equilibrium is the choice of Skiscim and Golden [SKI83]: define an *epoch* as a number of transitions with a fixed number of acceptances and the cost of an epoch as the cost value of the last configuration of an epoch. As soon as the cost of an epoch is within a specified distance from the cost of one of the preceding epochs, the Markov chain is terminated. Thus, termination of a Markov chain and, consequently, its length are related to fluctuations in the value of the cost function observed for that chain.
A similar proposal is made by Catthoor *et al.* [CAT85], who relate the length of the Markov chain to convergence of the denominator of eq. 3.64 and by Černy [ČER85], who does not quantify his proposal.

- decrement of the control parameter

As mentioned before, the decrement is chosen such that small Markov chain lengths suffice to re-establish quasi-equilibrium after the decrement of the control parameter. In section 5.1 it is argued that this leads to a decrement rule allowing only small changes in the value of the control parameter. A frequently used decrement rule is given by

$$c_{k+1} = \alpha \cdot c_k, \; k = 0, 1, 2, \ldots, \tag{5.8}$$

where α is a constant smaller than but close to 1. Kirkpatrick *et al.* are the first to propose this rule [KIR82], with $\alpha = 0.95$, but it is also widely used by others (Johnson *et al.* [JOHN87], Bonomi and Lutton [BON84], [BON86], [LUT86], Burkard and Rendl [BUR84], Leong *et al.* [LEO85a], [LEO85b], Morgenstern and Shapiro [MORC86], and

Sechen and Sangiovanni-Vincentelli [SEC85]), with values of α ranging from 0.5 to 0.99.
In [NAH85], Nahar *et al.* fix the number of decrement steps K in the value of the control parameter and values for c_k, $k = 1, \ldots, K$, are determined experimentally. Skiscim and Golden [SKI83], [GOLDE86] divide the interval $[0, c_0]$ into a fixed number of, say K, subintervals and c_k, $k = 1, \ldots, K$, is chosen as

$$c_k = \frac{K-k}{K} \cdot c_0, \; k = 1, \ldots, K, \tag{5.9}$$

i.e. instead of a constant ratio $\frac{c_{k+1}}{c_k}$ (see eq. 5.8), the difference between successive values of the control parameter is kept constant.

5.3 More elaborate cooling schedules

In this section some elaborate proposals for the choice of the parameters of a cooling schedule are discussed. We are well aware of the vast number of different proposals in the literature, but we consider the material presented in this section to be sufficiently representative in order to bring out the important aspects concerning cooling schedules.
First, the guidelines of White [WHI84] for choosing initial and final values of the control parameter are discussed as well as the proposals for the final value of Lundy and Mees [LUN86], Aarts and Van Laarhoven [AAR85a], Otten and Van Ginneken [OTT84] and Huang *et al.* [HUA86]. Next, the proposals of Romeo *et al.* [ROM85], [ROM84] and Huang *et al.* [HUA86] for the length of Markov chains are discussed. These proposals attempt to quantify the concept of quasi-equilibrium as discussed in the previous sections. Finally, we discuss the approaches by Aarts and Van Laarhoven [AAR85a], Lundy and Mees [LUN86], Otten and Van Ginneken [OTT84] and Huang *et al.* [HUA86], who relate the decrement rule for the control parameter to the quasi-equilibrium concept.

- initial value of the control parameter

The approach of White [WHI84] is based on the configuration density function, $\omega(C)$, given by eq. 4.21.
Assuming that $\omega(C)$ is *Gaussian* with mean $\overline{C}$ and standard deviation σ_∞, i.e. (see section 4.3)

$$\omega(C) = \exp\left(\frac{-(C-\overline{C})^2}{2{\sigma_\infty}^2}\right), \tag{5.10}$$

eq. 4.23 becomes [WHI84]

$$\langle C(c)\rangle = C_{opt} + \sqrt{2}\cdot\sigma_\infty\cdot\left(y + \frac{\exp(-y^2)}{\sqrt{\pi}(1+\mathrm{erf}(y))}\right), \tag{5.11}$$

where

$$y = \frac{1}{\sqrt{2}}\left(\frac{\overline{C}-C_{opt}}{\sigma_\infty} - \frac{\sigma_\infty}{c}\right). \tag{5.12}$$

For c large, eqs. 5.11 and 5.12 yield (see also eq. 4.29)

$$\langle C(c)\rangle \simeq \overline{C} - \frac{{\sigma_\infty}^2}{c}, \tag{5.13}$$

which is in accordance with the fact that for $c \to \infty$, $\langle C(c)\rangle$ approaches $\overline{C}$ (for $c \to \infty$, all configurations have equal probability of occurrence).
White now proposes c_0 be taken such that $\langle C(c_0)\rangle$ is just within one standard deviation ('thermal noise' [WHI84]) of $\overline{C}$. Using eq. 5.13, the requirement that $\overline{C} - \langle C(c_0)\rangle \leq \sigma_\infty$ implies that

$$c_0 \geq \sigma_\infty. \tag{5.14}$$

Approximating σ_∞ by using eq. 4.5, i.e. assuming that for large values of c,

$${\sigma_\infty}^2 \simeq \langle C^2(c)\rangle - (\langle C(c)\rangle)^2, \tag{5.15}$$

then yields

$$c_0 \geq \sqrt{\langle C^2(\infty)\rangle - (\langle C(\infty)\rangle)^2}, \tag{5.16}$$

where estimates for $\langle C^2(\infty)\rangle$ and $\langle C(\infty)\rangle$ are obtained by monitoring the evolution of the algorithm for the case that all transitions are accepted ($c = \infty$).

- final value of the control parameter

An expression for the final value of the control parameter is derived by White by considering a local minimum with cost value C_0. If C_1 is the next highest cost value of a configuration belonging to the neighbourhood of the local minimum, then White argues that low values of c should be such that

$$\exp\left(-\frac{(C_1 - C_0)}{c}\right) < R^{-1}, \tag{5.17}$$

i.e. such that the probability of accepting a transition from the local minimum to its next highest neighbour is smaller than R^{-1} (R is the size of the neighbourhoods), in order to guarantee $\langle C(c)\rangle < C_1$. Eq. 5.17 yields

$$c_f \leq \frac{(C_1 - C_0)}{\ln R}. \tag{5.18}$$

A sharper bound for c_f is obtained by Lundy and Mees in [LUN86] by requiring that, for the final value of the control parameter, in equilibrium (once the stationary distribution is reached) the probability for the current configuration of the Markov chain to be more than ϵ above the minimum value of the cost function should be less than some small real number ϑ, i.e. it is required that

$$Pr\{\mathbf{X}(k) = i \wedge C(i) > C_{opt} + \epsilon \mid c = c_f\} < \vartheta. \tag{5.19}$$

Using the fact that in equilibrium, $Pr\{\mathbf{X}(k) = i\} = q_i(c)$ is given by eq. 3.64, the following bound can be derived [LUN86]:

$$Pr\{\mathbf{X}(k) = i \wedge C(i) > C_{opt} + \epsilon\} \simeq$$

$$\sum_{i:C(i)>C_{opt}+\epsilon} q_i(c) < (|\mathcal{R}| - 1)\exp\left(-\frac{\epsilon}{c}\right) \tag{5.20}$$

and combining eqs. 5.19 and 5.20 yields the following condition on c_f

$$c_f \leq \frac{\epsilon}{\ln(|\mathcal{R}| - 1) - \ln\vartheta}. \tag{5.21}$$

Aarts and Van Laarhoven [AAR85a] as well as Otten and Van Ginneken [OTT84] propose a final value of the control parameter based on the decrease during execution of the algorithm of the average values of the cost function over a number of Markov chains. Let $\overline{C}(c_k)$ denote this average cost value observed at the k-th Markov chain, then execution of the algorithm is terminated once $\overline{C}(c_k) - C_{opt}$ is small. If $c_k \ll 1$ one obtains [AAR85a]

$$\overline{C}(c_k) - C_{opt} \simeq c_k \left(\frac{\partial \overline{C}(c)}{\partial c} \right)_{c=c_k}, \tag{5.22}$$

whence c_f is taken such that [AAR85a]

$$\left(\frac{\partial \overline{C}(c)}{\partial c} \right)_{c=c_f} \cdot \frac{c_f}{\overline{C}(c_0)} < \epsilon \tag{5.23}$$

for some small real number ϵ, or [OTT84]

$$\left(\frac{\partial \overline{C}(c)}{\partial c} \right)_{c=c_f} \cdot \frac{c_f}{(\overline{C}(c_0) - \overline{C}(c_f))} < \epsilon. \tag{5.24}$$

Moreover, Otten and Van Ginneken use eqs. 4.5 and 4.7 to rewrite eq. 5.24 as

$$\frac{\overline{C}^2(c_f) - (\overline{C}(c_f))^2}{c_f \cdot (\overline{C}(c_0) - \overline{C}(c_f))} < \epsilon. \tag{5.25}$$

Finally, Huang *et al.* [HUA86] propose the following stop criterion: at the end of each Markov chain, compare the difference between the maximum and minimum cost values among the accepted transitions for that chain with the maximum change in cost for any transition accepted during the generation of that chain. If they are the same, all configurations in the current Markov chain are of about the same cost and there is no need to use simulated annealing: c is set to 0 and the optimization is concluded with a local search (iterative improvement).

- length of Markov chains

Exact quantification of the quasi-equilibrium concept is an insurmountable task (the major problem being to collect sufficient statistics

to determine accurately the probability distribution of the configurations).
Romeo and Sangiovanni-Vincentelli [ROM85] observe that, in order to obtain a final configuration close to a globally minimal one, there should always be a sufficiently large probability to leave any configuration, possibly a local minimum, found during execution of the algorithm. A straightforward calculation shows that for any configuration i the expected number of transitions to leave $i, \tilde{N}_i(c)$, is given by [ROM85]:

$$\tilde{N}_i(c) = (1 - P_{ii}(c))^{-1}. \tag{5.26}$$

Since $P_{ii}(c)$ is given by

$$P_{ii}(c) = 1 - \frac{1}{R} \sum_{j \in \mathcal{R}_i} \min\{1, \exp(-(C(j) - C(i))/c)\}, \tag{5.27}$$

evaluation of $P_{ii}(c)$, for **all** $i \in \mathcal{R}$, requires the values of the cost function of all configurations. Since one is only interested in the maximum value $\tilde{N}(c) \stackrel{\text{def}}{=} \max_{i \in \mathcal{R}} \tilde{N}_i(c)$, or, equivalently, the maximum value of $P_{ii}(c)$ (over all $i \in \mathcal{R}$), a conservative estimate is obtained by approximating $C(j) - C(i)$ (for all i and j) by $C(\hat{j}) - C(\hat{i})$, where $\hat{i}$ and $\hat{j}$ are the configurations with lowest and highest value of the cost function, respectively, obtained so far during execution of the algorithm.
Thus, the following estimate for $\tilde{N}(c)$ is obtained

$$\tilde{N}(c) \simeq (\exp(-(C(\hat{j}) - C(\hat{i}))/c))^{-1}. \tag{5.28}$$

Romeo and Sangiovanni-Vincentelli now propose that the Markov chain length be taken proportional to $\tilde{N}(c)$. Further motivation for this choice is given in [ROM84] by observing that in a Markov chain, viewed as a *stochastic dynamical* system, $\tilde{N}(c)$ plays a role similar to the time constant in *linear dynamical* systems. In such a system, when controlled by a piece-wise constant function, a time as long as a few time constants is usually sufficient to bring the system to a new equilibrium.
Huang *et al.* [HUA86] base the length of the Markov chains on the observation that in equilibrium the ratio of the number of new configurations generated with their costs within a certain range δ from the

average cost to the total number of generated and accepted configurations reaches a stationary value κ. Assuming a normal distribution for the values of the cost function, it can be shown that $\kappa = \mathrm{erf}\left(\frac{\delta}{\sigma(c)}\right)$ [HUA86]. Huang *et al.* define the *within count* and the *tolerance count* as the number of accepted configurations with a cost value inside and outside the interval $(\langle C(c)\rangle - \delta, \langle C(c)\rangle + \delta)$, respectively, and consider equilibrium to be reached if the within count exceeds a target value (typically $p \cdot \kappa$, where p is some parameter depending on the size of the problem) before the tolerance count exceeds a maximum value (given by $p \cdot (1 - \kappa)$). Some extra conditions are imposed to avoid extremely long Markov chains (for low values of c, for example, the target value may never be reached, since virtually no configurations are accepted).

- decrement rule for the control parameter

As was already mentioned, the decrements are chosen small in order to avoid the necessity of long Markov chains for re-establishing equilibrium at each new value of the control parameter. Thus, Aarts and Van Laarhoven [AAR85a], Lundy and Mees [LUN86] and Otten and Van Ginneken [OTT84] argue that the stationary distributions for two succeeding values of the control parameter should be close, i.e.

$$\forall i \in \mathcal{R} : \frac{1}{1+\delta} < \frac{q_i(c_k)}{q_i(c_{k+1})} < 1 + \delta,\ k = 0, 1, \ldots, \tag{5.29}$$

for some small real number δ. If eq. 5.29 is satisfied, then the probability distribution of the configurations should, after changing c_k into c_{k+1}, rapidly approach the new stationary distribution $\mathbf{q}(c_{k+1})$, provided there was quasi-equilibrium at c_k.

The three aforementioned sets of authors all take eq. 5.29 as a starting point but derive alternative decrement rules:

1. Aarts and Van Laarhoven [AAR85a] show that eq. 5.29 is satisfied if

$$\forall i \in \mathcal{R} : \exp\left(-\frac{(C(i) - C_{opt})}{c_k}\right) < (1+\delta) \cdot \exp\left(-\frac{(C(i) - C_{opt})}{c_{k+1}}\right). \tag{5.30}$$

Assuming that the values of the cost function for the k-th Markov chain are normally distributed with mean $\mu(c_k) = \overline{C}(c_k) -$

C_{opt} and variance $\sigma^2(c_k) = \overline{C}^2(c_k) - (\overline{C}(c_k))^2$ (see section 4.3, postulate 1), it is shown that eq. 5.30 is equivalent to

$$c_{k+1} > c_k \cdot \left(1 + \frac{\ln(1+\delta) \cdot c_k}{\mu(c_k) + 3\sigma(c_k)}\right)^{-1}, \tag{5.31}$$

which yields the following decrement rule:[1]

$$c_{k+1} = c_k \cdot \left(1 + \frac{\ln(1+\delta) \cdot c_k}{3\sigma(c_k)}\right)^{-1}, \tag{5.32}$$

where it is argued that the neglect of $\mu(c_k)$ can be taken into account by choosing smaller values of δ, since $\mu(c_k)$ and $\sigma(c_k)$ have about the same dependence on c_k [AAR88a].

2. Lundy and Mees [LUN86] show that eq. 5.29 leads to

$$\forall i \in \mathcal{R} : \exp\left(\frac{(C(i) - C_{opt})(c_k - c_{k+1})}{c_k \cdot c_{k+1}}\right) < 1 + \delta, \tag{5.33}$$

which results in

$$\frac{c_k - c_{k+1}}{c_k \cdot c_{k+1}} \ll \frac{1}{U}, \tag{5.34}$$

where U is some upper bound on $(C(i) - C_{opt})$. Choosing

$$\frac{c_k - c_{k+1}}{c_k \cdot c_{k+1}} = \frac{\gamma}{U}, \tag{5.35}$$

for some small real number γ then yields the following decrement rule:

$$c_{k+1} = c_k \cdot \left(1 + \frac{\gamma \cdot c_k}{U}\right)^{-1}, \tag{5.36}$$

which is similar to the one given by eq. 5.32.

---

[1] The decrement rule proposed by Lam and Delosme [LAM86b] is almost identical; the ratio $\frac{c_k}{\sigma(c_k)}$ in eq. 5.32 is replaced by $\left(\frac{c_k}{\sigma(c_k)}\right)^3$.

3. Otten and Van Ginneken's ([OTT84]) derivation is partly based on eq. 5.29 and partly on their experimental observation that the best results in the trade-off between fast decrement and small Markov chain length (see the discussion in section 5.1) are obtained if the decrements in c are kept inversely proportional to the changes in entropy (as defined by eq. 4.6). The latter observation leads to a decrement in c, proportional to

$$\frac{{c_k}^3}{(\sigma(c_k))^2} \tag{5.37}$$

(cf. eq. 4.15), whereas eq. 5.29 is claimed to be equivalent to

$$| c_k - c_{k+1} | \leq \frac{{c_k}^2 \cdot \ln(1+\delta)}{C_{max} + c_k \cdot \ln(1+\delta)}. \tag{5.38}$$

Combining eqs. 5.37 and 5.38 yields the following decrement rule [OTT84]

$$c_{k+1} = c_k - \frac{1}{M_k} \cdot \frac{{c_k}^3}{(\sigma(c_k))^2}, \tag{5.39}$$

where M_k is given by

$$M_k = \frac{C_{max} + c_k \cdot \ln(1+\delta)}{\sigma(c_k)^2 \cdot \ln(1+\delta)} \cdot c_k. \tag{5.40}$$

Otten and Van Ginneken remark that the decrement in c, as given by eq. 5.39, may be too fast. As an additional control parameter, they use the entropy $S(c)$, defined by eq. 4.6. Since $\lim_{c \downarrow 0} S(c) = \ln | \mathcal{R}_{opt} |$ (see eq. 4.13), a rule to check the "quality" of the final configuration obtained by the algorithm at c_f can be based on the quantity $\Xi(c_f)$, given by

$$\Xi(c_f) = \left| \frac{S(c_f) - \ln | \mathcal{R}_{opt} |}{S(c_0) - \ln | \mathcal{R}_{opt} |} \right|. \tag{5.41}$$

Otten and Van Ginneken approximate $S(c)$ and $\ln | \mathcal{R}_{opt} |$ and propose that execution of the algorithm be repeated with a decrement rule given by

$$c_{k+1} = c_k - \frac{1}{2M_k} \cdot \frac{{c_k}^3}{(\sigma(c_k))^2}, \tag{5.42}$$

if, after termination of the algorithm, $\Xi(c_f) > \eta$, for some small real number η. This division of the decrement in c into halves and re-execution of the algorithm is repeated until finally $\Xi(c_f) < \eta$. Aarts and Van Laarhoven [AAR85a], however, doubt that a condition for re-execution of the algorithm can be based on eq. 5.41 using the small Markov chain length proposed by the authors, since their experimental observations indicate that the entropy can only be approximated with sufficient accuracy by using long Markov chains.

Huang *et al.* [HUA86] base the decrement rule on the average cost values of consecutive Markov chains. Let $\overline{C}(c_k)$ denote again the average cost value observed at the k-th Markov chain. For the expected cost value in equilibrium, $\langle C(c)\rangle$, defined by eq. 4.3, we have the following relation (cf. eq. 4.7)

$$\frac{\partial}{\partial \ln c}\langle C(c)\rangle = \frac{\sigma^2(c)}{c}. \tag{5.43}$$

Approximating $\langle C(c_k)\rangle$ by $\overline{C}(c_k)$, we find

$$\frac{\overline{C}(c_{k+1}) - \overline{C}(c_k)}{\ln c_{k+1} - \ln c_k} \simeq \frac{\sigma^2(c_k)}{c_{k+1}}, \tag{5.44}$$

which leads to

$$c_{k+1} = c_k \exp\left(\frac{c_k(\overline{C}(c_{k+1}) - \overline{C}(c_k))}{\sigma^2(c_k)}\right). \tag{5.45}$$

To maintain quasi-equilibrium, Huang *et al.* require the difference in average cost for two consecutive Markov chains to be less than the standard deviation of the cost: $\overline{C}(c_{k+1}) - \overline{C}(c_k) = -\lambda\sigma(c_k)$, where $\lambda \leq 1$. This results in the following decrement rule:

$$c_{k+1} = c_k \exp\left(-\frac{\lambda c_k}{\sigma(c_k)}\right). \tag{5.46}$$

The various proposals discussed in the present section are schematically depicted in table 5.1. Using this table as a guideline, the characteristic features of cooling schedules used in the simulated annealing algorithm can be summarized as follows. With respect to the **initial value of the control parameter**, all proposed choices are similar and based on the argument that initially virtually all transitions are to be accepted. However, there is a clear dichotomy in approaches to the choice of a **decrement rule** for c and the **length of Markov chains** L. The following two general classes can be distinguished:

- Class A: a variable Markov chain length and a fixed decrement of the control parameter, and
- Class B: a fixed Markov chain length and a variable decrement of the control parameter.

Here, "fixed" and "variable" refer to independence and dependence, respectively, on the evolution of the algorithm. The subdivision into two classes A and B almost coincides with the distinction between conceptually simple cooling schedules and more elaborate ones, Romeo and Sangiovanni-Vincentelli's as well as the schedule of Huang *et al.* being the only exceptions since they combine a variable decrement rule with a variable Markov chain length.
With respect to the last parameter (**final value** of c) two approaches can be distinguished: in conceptually simple cooling schedules, execution of the algorithm is terminated if no transitions are accepted during a number of Markov chains, whereas more elaborate choices are based on extrapolations of the average cost of configurations over a number of consecutive Markov chains.
The performance of the simulated annealing algorithm with different cooling schedules is briefly discussed in the next chapter.

5.4 Improvements of the generation mechanism for transitions

This chapter is concluded with the discussion of two approaches related to implementation aspects of the simulated annealing algorithm. Both approaches are based on improvements of the generation mechanism for transitions, resulting in a faster execution of the algorithm.

During execution of the simulated annealing algorithm, the value of

class	*schedule*	*init. value of c*	*decrement in c*
A	'simple' (Sect. 5.2)	$c_0 = \frac{\Delta C^{(+)}}{\ln(\chi_0^{-1})}$ (eq. 5.7)	$c_{k+1} = \alpha \cdot c_k$ $\alpha \in [0.5, 0.99]$
	[ROM85]	-	$c_{k+1} = \alpha \cdot c_k$ α between 0.8-0.99
	[HUA86]	identical to [WHI84] (eq. 5.14)	$c_{k+1} =$ $c_k \cdot \exp\left(-\frac{\lambda c_k}{\sigma(c_k)}\right)$ (eq. 5.46)
	[AAR85a]	similar to eq. 5.7	$c_{k+1} =$ $c_k \cdot \left(1 + \frac{\ln(1+\delta)\cdot c_k}{3\sigma(c_k)}\right)^{-1}$ (eq. 5.32)
B	[LUN86]	"	$c_{k+1} =$ $c_k \cdot (1 + \frac{\gamma \cdot c_k}{U})^{-1}$ (eq. 5.36)
	[OTT84]	"	$c_{k+1} =$ $c_k - \frac{1}{M_k} \cdot \frac{c_k^3}{(\sigma(c_k))^2}$ (eq. 5.39)
class	*schedule*	*final value of c*	*chain length L*
A	'simple' (Sect. 5.2)	no change in cost over a fixed no. of Markov chains	fixed or determined by minimum no. of accepted transitions (with upperb.)
	[ROM85]	"	proportional to $\tilde{N}$ (eq. 5.28)
	[HUA86]	see text	based on $\kappa = \mathrm{erf}\left(\frac{\delta}{\sigma(c_k)}\right)$ (see text)
	[AAR85a]	$c_f \le \frac{\epsilon \overline{C}(c_0)}{(\partial \overline{C}(c)/\partial c)_{c=c_f}}$ (eq. 5.23)	$R = \mid \mathcal{R}_i \mid$
B	[LUN86]	$c_f \le \frac{\epsilon}{\ln(\mid\mathcal{R}\mid - 1) - \ln \vartheta}$ (eq. 5.21)	-
	[OTT84]	$c_f \le \frac{\epsilon(\overline{C}(c_0) - \overline{C}(c_f))}{(\partial \overline{C}(c)/\partial c)_{c=c_f}}$ (eq. 5.25)	proportional to the number of of variables

Table 5.1: Summary of discussed cooling schedules

the control parameter is gradually decreased and the lower the value of c, the lower the probability for a transition, corresponding to a large increase in the cost function, to be accepted. However, in [WHI84], White argues that these transitions can often help the system approach equilibrium faster. For high values of c, where virtually all transitions are accepted, it would therefore be helpful to bias the generation of transitions in favour of 'large' transitions.

The biased generation of transitions is based on the concept of *transition classes* (in White's terminology they are called *move classes*). The class of transitions of length 1 consists of all transitions $i \rightarrow j$, for which $G_{ij} > 0$. If two configurations i and j are connected by at least l transitions of length 1, then a transformation of i in j is defined as a transition of length l. White remarks that a single transition of length l can often be implemented at less computational cost than l transitions of length 1.

At high values of c, large transitions are favoured since they result in a faster convergence to equilibrium. As c decreases, the generation of transitions is restricted to those with a smaller length (large transitions would not be accepted anyhow).

Finally, White claims that for each value of c, the transitions should be such that $\langle | \Delta C | \rangle$ is somewhat less than the average fluctuation around $\langle C(c) \rangle$, where ΔC is the change in cost as the result of a transition. Thus, at any value of c, transition classes should be selected for which

$$\langle | \Delta C | \rangle \simeq c \tag{5.47}$$

and

$$\langle | \Delta C | \rangle < \sqrt{\langle C^2(c) \rangle - (\langle C(c) \rangle)^2}. \tag{5.48}$$

A typical feature of the simulated annealing algorithm is that for low values of c, generated transitions are often rejected, since they correspond to (possibly large) increases in the cost function, which are accepted with low probabilities. Thus, if a cooling schedule is used which requires a minimal number of accepted transitions for each value of c (see section 5.2), the length of Markov chains and consequently the computation time increase as c decreases.

Greene and Supowit [GREE84] propose to bias the generation of transitions by using a list of the effects of each possible transition on the cost function. Suppose that configuration i is given and let $W_{ij}(c)$ for each of the R possible transitions be given by

$$W_{ij}(c) = \min\{1, \exp(-(C(j) - C(i))/c)\}, \tag{5.49}$$

where c is the current value of the control parameter. Instead of the traditional generation, where G_{ij} is given by

$$G_{ij} = R^{-1}, \tag{5.50}$$

$G_{ij}(c)$ is now defined as

$$G_{ij}(c) = \frac{W_{ij}(c)}{\sum_{k=1}^{R} W_{ik}(c)} \quad \forall j \in \mathcal{R}_i \tag{5.51}$$

and $A_{ij} = 1$ (for all i and j), i.e. all transitions are accepted once they are generated. Thus, if the length of a Markov chain is determined by a minimal number of accepted transitions Υ, then each Markov chain will have length Υ. Greene and Supowit, therefore, introduce the notion *rejectionless method* for their proposal.
By choosing the generation matrix $G(c)$ according to eq. 5.51, the sequence of configurations $\{\mathbf{X}(k)\}$, $k = 0, 1, \ldots$, generated by the rejectionless method is probabilistically equivalent to the sequence generated by the simulated annealing algorithm in its original formulation. This is shown by Greene and Supowit by considering the probability that in the original algorithm, possibly after a number of rejected transitions, a configuration i is exited (to a configuration $j \in \mathcal{R}_i$). This probability, for a given value of the control parameter c, is given by

$$\sum_{k=0}^{\infty} (P_{ii}(c))^k P_{ij}(c) = \sum_{k=0}^{\infty} (P_{ii}(c))^k \frac{1}{R} \cdot W_{ij}(c) =$$

$$\frac{R^{-1} \cdot W_{ij}(c)}{1 - P_{ii}(c)} = \frac{W_{ij}(c)}{\sum_{k=1}^{R} W_{ik}(c)}, \tag{5.52}$$

which is the probability of generating the transition $i \to j$ in the rejectionless method.
Moreover, Greene and Supowit claim that the stationary distribution

$\mathbf{q}'(c)$ of the Markov chain generated according to eq. 5.51 equals the stationary distribution given by eq. 3.40, if each component $q_i(c)$ is weighted by the expected number of transitions to leave i, $\tilde{N}_i$ (given by eq. 5.26), i.e.

$$\frac{q_i'(c)\tilde{N}_i}{\sum_{j\in\mathcal{R}} q_j'(c)\tilde{N}_j} = \frac{\exp(-\frac{C(i)-C_{opt}}{c})}{\sum_{j\in\mathcal{R}}\exp(-\frac{C(j)-C_{opt}}{c})}. \tag{5.53}$$

Suppose that the current configuration is i_1. First, G_{i_1j} is calculated for all $j \in \mathcal{R}_{i_1}$ and some $i_2 \in \mathcal{R}_{i_1}$ is accepted. Next, G_{i_2k} must be calculated for all $k \in \mathcal{R}_{i_2}$ etc., i.e. the rejectionless method requires R calculations (of the G_{ij}'s) for **each** transition. However, for the application discussed by Greene and Supowit [GREE84] (the *net partitioning problem*, a generalization of the *graph partitioning problem*, see also section 6.4), there is a simple relation between the successive neighbourhoods $\mathcal{R}_{i_1}$ and $\mathcal{R}_{i_2}$, giving an explicit expression for $\mathcal{R}_{i_2}\setminus\mathcal{R}_{i_1}$. Using this relation, only those G_{i_2k}'s have to be calculated for which $k \in \mathcal{R}_{i_2} \setminus \mathcal{R}_{i_1}$ (generally, $|\mathcal{R}_{i_2} \setminus \mathcal{R}_{i_1}|$ is small compared with R).
In general, however, it is unclear how to avoid evaluation of *all* G_{ij}'s after *each* accepted transition if there is no such simple relation between neighbourhoods of successive configurations. Such a relation being available, the advantage of the rejectionless method is clear: in a cooling schedule, where the Markov chain length is based on a minimal number of transitions, all Markov chains can have equal length.
As an aside, Greene and Supowit consider an alternative choice for $W(c)$, given by [GREE86]

$$W_{ij}(c) = \exp\left(-\frac{C(j)-C(i)}{2c}\right) \tag{5.54}$$

and from preliminary results they conclude that, at a given value of the control parameter, this choice can lead to a slightly faster convergence to equilibrium than the original choice (eq. 5.49).

Chapter 6

Performance of the simulated annealing algorithm

6.1 Introduction

The *performance analysis* of an approximation algorithm concentrates on the following two quantities:

- the **quality** of the final solution obtained by the algorithm, i.e. the difference in cost value between the final solution and a globally minimal configuration;
- the **running time** required by the algorithm.

For the simulated annealing algorithm, these quantities depend on the problem instance as well as the cooling schedule.
Traditionally, i.e. for deterministic algorithms, one distinguishes between two different types of performance analysis, *worst-case* analysis and *average-case* analysis. The worst-case analysis is concerned with upper bounds on quality and running time, the average-case analysis with expected values of quality and running time for a given probability distribution of problem instances. Since the simulated annealing algorithm is a *probabilistic* algorithm, an additional probabilistic aspect is added to the aforementioned classification. Besides the probability distribution over the set of problem instances, there is also a

probability distribution over the set of possible solutions, which the algorithm can obtain for a given problem. Thus, in an average-case analysis the average can refer to either the set of problem instances or the set of solutions of a given problem instance.
Investigating the performance of the simulated annealing algorithm, one might think of two approaches:

- a *theoretical* analysis: trying to find analytical expressions for the quality of the solution as well as for the running time, given a particular problem instance and cooling schedule.

- an *empirical* analysis: solving many instances of different combinatorial optimization problems with different cooling schedules and drawing conclusions from the results, with respect to both quality and running time;

As far as the theoretical analysis is concerned, the literature provides only worst-case analysis results (see section 6.2):

- it can be shown that some cooling schedules can be executed in a number of elementary operations bounded by a polynomial in the problem size;

- upper bounds can be given for the proximity of the probability distribution of the configurations after generation of a finite number of transitions to the uniform probability distribution on the set of optimal configurations (the asymptotic probability distribution, see sections 3.1 and 3.2). An upper bound on the quality of the final solution is only known for the *maximum matching problem* (a problem which is in P).

To the best of our knowledge, no theoretical average-case analysis results are known for the running time and quality of solution.
Section 6.3 is devoted to the work of Bonomi and Lutton ([BON84], [BON86], [LUT86]). This work concentrates on the performance of the simulated annealing algorithm on large, randomly generated, combinatorial optimization problems. These authors also address problems in *probabilistic value analysis* [LEN82], the probabilistic description of the optimal solution of a combinatorial optimization problem as a

function of problem parameters.
Empirical analysis results are reported by a number of authors for various combinatorial optimization problems (section 6.4). Most of these results focus on the quality of the final solution and the corresponding running time obtained by solving a given problem instance. More systematic analysis results for the performance of the algorithm with a particular cooling schedule are presented by Aarts *et al.* [AAR88a] and Randelman and Grest [RAN86] (see section 6.4). For applications of the algorithm to problems where the emphasis is on **solving the problem**, rather than on **analyzing the performance**, the reader is referred to chapter 7.

6.2 Worst-case performance results

Both Lundy and Mees [LUN86] and Aarts and Van Laarhoven [AAR85a] have shown that, using their decrement rules, given by eqs. 5.36 and 5.32, respectively, as well as their stop criteria (final values of the control parameter given by eqs. 5.21 and 5.23, respectively) it is possible to derive an upper bound for the number of steps in the value of the control parameter.
The bound for this number, s, derived by Lundy and Mees [LUN86] is given by

$$s < \frac{(\ln |\mathcal{R}| - \ln \alpha) \cdot U}{\gamma \cdot \epsilon}, \tag{6.1}$$

where α, γ, ϵ and U are as in eqs. 5.21 and 5.36, respectively. Aarts and Van Laarhoven's bound [AAR85a] is given by

$$s < \frac{\ln |\mathcal{R}|}{\alpha \cdot \epsilon' \cdot \overline{C}(c_0)}, \tag{6.2}$$

for some $\epsilon' \in (0, \epsilon]$ and α is given by

$$\alpha = \min_k \left(\frac{\ln(1+\delta)}{3\sigma(c_k)} \right). \tag{6.3}$$

Thus, both bounds imply that the number of steps in the value of the control parameter is $\mathcal{O}(\ln |\mathcal{R}|)$. Combined with a fixed Markov chain length (as in [AAR85a], where $L = R$, the size of the neighbourhoods), this yields the following worst-case result for the total number of transitions generated during execution of the algorithm, s_{tot}:

$$s_{tot} = \mathcal{O}(R \cdot \ln |\mathcal{R}|). \tag{6.4}$$

Since for most combinatorial optimization problems, the sizes of the neighbourhoods can be chosen polynomial and the size of the set of configurations, $|\mathcal{R}|$, is exponential in the size of the input of the problem [PAP82], eq. 6.4 yields that execution of the algorithm takes polynomial time for most combinatorial optimization problems, which is, in view of the discussion in chapter 1, a satisfactory result.
Of course, one might object that such a result is worthless without some guarantee for the proximity of the final configuration to an optimal one. To the best of our knowledge, the only upper bound for this proximity is a probabilistic one, given by Sasaki and Hajek [SAS88] for the *maximum matching* problem: given a graph $G = (V, E)$, find a *matching* M of maximum cardinality, where a matching is defined as a subset of E such that no two edges in M are incident to the same vertex. Let $\mathbf{X}(k)$ be the matching obtained after k transitions of the annealing algorithm and let the transition mechanism be defined as follows: an edge $e \in E$ is chosen at random. If $e \notin \mathbf{X}(k)$ and $\mathbf{X}(k) \cup \{e\}$ is a matching then $\mathbf{X}(k+1) = \mathbf{X}(k) \cup \{e\}$, if $e \in \mathbf{X}(k)$ then delete e from $\mathbf{X}(k)$ with probability $\exp\left(-\frac{1}{c_k}\right)$, else set $\mathbf{X}(k+1) = \mathbf{X}(k)$. Sasaki and Hajek prove that for any number $\beta > 1$ and a particular choice of c_k (independent of k!), the expected value $\langle \underline{R} \rangle$ satisfies

$$\langle \underline{R} \rangle \leq 24(1+\beta)^2 |V|^5 (2d^*)^{2\beta}, \tag{6.5}$$

where $\underline{R}$ is a random variable denoting the first transition r leading to a matching $\mathbf{X}(r)$ satisfying

$$|\mathbf{X}(r)| \geq \lfloor C_{opt} \left(1 - \frac{1}{\beta}\right) \rfloor, \tag{6.6}$$

and d^* denotes the maximum degree of the vertices of the graph. Thus, the expected time to find a matching deviating less than $\frac{100}{\beta}\%$ from the global minimum can be bounded by a polynomial in the problem size. At the same time, however, Sasaki and Hajek show

that for a special class of graphs the expected time to find an optimal configuration *cannot* be bounded by a polynomial in $|V|$.
A deterministic upper bound for the proximity of the probability distribution of the configurations after generation of k transitions to the uniform probability distribution on the set of optimal configurations has been obtained independently by Anily and Federgruen [ANI87a], Gidas [GID85a] and Mitra *et al.* [MIT86]. All bounds are derived for the inhomogeneous algorithm (see section 3.2) and lead to similar worst-case complexities. As an example the result of Mitra *et al.* [MIT86] is discussed.
Consider the inhomogeneous algorithm (decrement of c after each transition) and let the sequence $\{c_k\}$, $k = 0, 1, \ldots$ of values of the control parameter be such that the equality in eq. 3.74 is satisfied. If $\mathbf{a}(c_k)$ is the probability distribution of the configurations after generation of k transitions and π the uniform distribution on the set of optimal configurations (given by eq. 3.5), then

$$\sum_{i=1}^{|\mathcal{R}|} | a_i(c_{kr}) - \pi_i | = \mathcal{O}\left(\frac{1}{k^{\min(a,b)}}\right), \tag{6.7}$$

where r is given by eq. 3.73 and a and b given by

$$a = \frac{1}{r}(\min_{i \in \mathcal{R}} \min_{j \in \mathcal{R}_i} G_{ij})^r \tag{6.8}$$

and

$$b = \frac{\delta}{r \cdot \Delta}, \tag{6.9}$$

respectively, where δ is the difference between C_{opt} and the next-to-least cost value and Δ is given by eq. 3.66. This bound is rather poor, however, in the sense that if one works it out for a particular problem one typically finds that the time required for good accuracy is larger than the number of configurations (for the n-city travelling salesman problem, for example, one finds that k is $\mathcal{O}\left(\varepsilon^{-n^{2n+1}}\right)$, where ε is the required accuracy, whereas the number of configurations is $\mathcal{O}(n!)$).
Another bound is obtained by Gelfand and Mitter [GEL85]. For a given subset I of $\mathcal{R}$ and a decrement rule of the form $c_k = \frac{\Gamma}{\log k}$, for some constant Γ, they obtain a lower bound on $Pr\{\mathbf{X}(n) \in I, n \leq k\}$,

i.e. the probability that the Markov chain visits a configuration in I at least once during the generation of k transitions. Taking $I = \mathcal{R}_{opt}$, a lower bound is obtained on the probability that the simulated annealing algorithm finds a global minimum at least once during the generation of k transitions. Gelfand and Mitter show that, if I is given by eq. 3.85, the bound converges exponentially fast to 1 (for $k \to \infty$), if Γ is large enough. For small values of Γ the bound converges to a value greater than 0 (and possibly smaller than 1).

6.3 Probabilistic value analysis

Bonomi and Lutton analyse the performance of the simulated annealing algorithm on three problems, *travelling salesman* [BON84], *matching* [LUT86] and *quadratic assignment* [BON86], by randomly generating instances of these problems and analyzing the behaviour of the algorithm as the problem size goes to infinity. The interest in this kind of analysis is twofold:

- by using some of the quantities, based on the analogy between combinatorial optimization and statistical mechanics (chapter 4), it is possible to derive in a straightforward way analytical expressions for $\frac{C_{opt}}{\sqrt{n}}$ in cases where the problem size goes to infinity;
- using the aforementioned expressions, it is possible to compare the cost value of the solution, obtained by simulated annealing, with the globally minimal value.

Hereinafter, these problems are presented separately and after a short description of each problem the corresponding results obtained by Bonomi and Lutton are discussed.

(1) the Travelling Salesman Problem (TSP) [LAW85]

In the n-city TSP, a distance matrix $D = (d_{ij}),\ i,j = 1,\ldots,n$ is given; d_{ij} denotes the distance between cities i and j. A tour through the n cities is defined as a closed walk that visits each city exactly once. Each tour can be represented by an element π of the set of all cyclic permutations of the n cities $\{1,\ldots,n\}$, if π is defined such that

$\pi(i)$, $i = 1, \ldots, n$ is the successor of city i in the tour. Thus, the set of configurations consists of all cyclic permutations of $\{1, \ldots, n\}$ (there are $\frac{1}{2}(n-1)!$ such permutations for a TSP with a symmetric distance matrix) and the cost of a permutation is defined as the length of the corresponding tour, i.e.

$$C(\pi) = \sum_{i=1}^{n} d_{i\pi(i)}. \tag{6.10}$$

The TSP now is to minimize the cost over all possible permutations. Consider the class of problem instances, where the n cities are points in the two-dimensional Euclidean space, whose locations are drawn independently from the uniform distribution over the unit square, and let d_{ij} be defined as the Euclidean distance between the locations of i and j. If $C_{opt}^{(D)}$ is the smallest tour length in an instance of this class with distance matrix D, then in a landmark paper Beardwood *et al.* [BEA59] have shown that

$$\lim_{n \to \infty} \frac{C_{opt}^{(D)}}{\sqrt{n}} = \theta, \tag{6.11}$$

where θ is an unknown constant (numerical estimates by means of *Monte Carlo experiments* yield $\theta \simeq 0.749$).
Bonomi and Lutton [BON84] apply the simulated annealing algorithm to the aforementioned class of TSPs and show that, as n goes to infinity, the cost value of the final configuration found by the algorithm indeed approaches $\theta\sqrt{n}$ within a few percent (with $\theta = 0.749$). In an earlier paper, Kirkpatrick *et al.* [KIR82] describe a similar experiment and report similar values for $\frac{C_{final}}{\sqrt{n}}$.
In order to be able to apply the simulated annealing algorithm a neighbourhood structure must be defined (cf. chapter 1), i.e. for each tour we have to define a neighbourhood as the set of tours that can be reached from the present tour in one transition. A well-known neighbourhood structure for TSP (which is, almost without exception, used in applications of simulated annealing to TSP, see section 6.4) is defined by the generation mechanism for transitions introduced by Lin and Kernighan [LIN65], [LIN73]. The corresponding transitions are known as *k-opt transitions.* The simplest case, a 2-opt transition, is

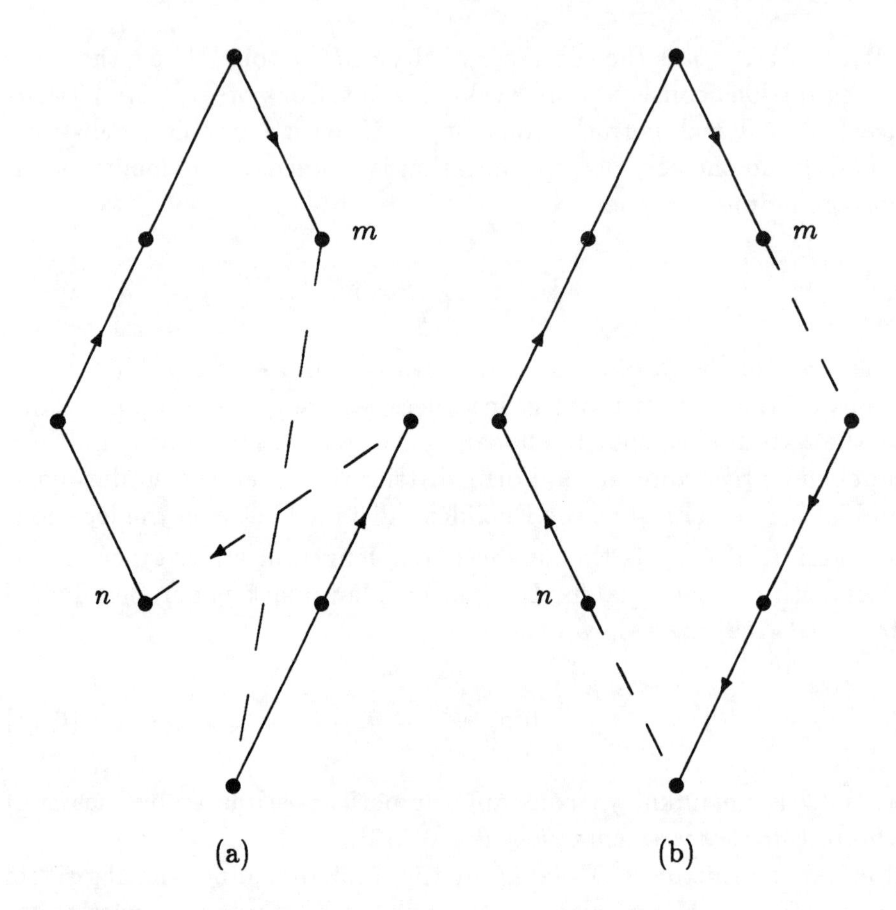

Figure 6.1: 2-opt example; (a) current tour (b) tour after reversing the order between cities m and n.

carried out by selecting two cities in the present tour and reversing the order in which the cities in between this pair of cities are visited (see figure 6.1). A neighbourhood of a tour is now defined as the set of all tours that can be reached from the current tour via a 2-opt transition (there are $\frac{1}{2}n(n-1)$ such neighbours).
Bonomi and Lutton's cooling schedule is of the type discussed in section 5.2: fixed decrement in c ($\alpha = 0.925$), Markov chain length proportional to n and a fixed number of values for the control parameter, viz. 50. Additionally, the generation mechanism is restricted in the sense that the unit square is divided into a number of subregions

(also proportional to n). The two cities are randomly chosen from $\{1, 2, \ldots, n\}$, but such that they either belong to the same subregion or to two adjacent ones. In this way the running time for each transition is $\mathcal{O}(n^{p'})$ where $p' < 1$ and thus the total running time for the algorithm is $\mathcal{O}(n^p)$ where $1 < p < 2$ (one n is contributed by the Markov chain length).
Two instances are presented, both belonging to the aforementioned class: a 400-city TSP, for which the cost value of the final solution is indeed $0.749 \cdot \sqrt{n}$, and a 10000-city TSP, for which $0.763 \cdot \sqrt{n}$ is found. Thus, the conjectured optimal solutions are approached within a few percent in better than quadratic time. Furthermore, these results compare favourably with the results obtained with Lin's 2-opt algorithm [LIN65] and the *convex hull* algorithm [GOLDE80], in terms of both quality of solution and running time.
Bonomi and Lutton also derive analytical expressions for the expected tour length and entropy in equilibrium (see section 4.2) for the aforementioned class of TSPs and it is shown that these expressions can be approximated quite accurately by applying the simulated annealing algorithm with the aforementioned cooling schedule to these problems.

(2) the Quadratic Assignment Problem (QAP)

In the QAP, two $n \times n$-matrices $D = (d_{ij})$ and $T = (t_{kl})$ are given. The set of configurations of the problem consists of all permutations of the rows and columns of T, $\{1, 2, \ldots, n\}$; the cost of each permutation π is given by

$$C(\pi) = \sum_{i=1}^{n} \sum_{j=1}^{n} d_{ij} t_{\pi(i)\pi(j)}. \tag{6.12}$$

The QAP is the problem of minimizing C over all $n!$ permutations and can be viewed as the problem of finding the best assignment of n sources to n sites, where d_{ij} is defined as the distance from site i to site j, t_{kl} as the amount of traffic which is to be routed from source k to source l and $\pi(i)$ as the source assigned to site i.
Consider the class of problem instances where the locations of the n sites are again uniformly distributed on the unit square, with average distance $\langle d \rangle$, and the matrix elements t_{kl} are independent random

variables as well, with expectation $\langle t\rangle$. Burkard and Fincke [BUR83a] have shown that for large n, the difference between worst and best assignment disappears, i.e.

$$\forall\pi : \lim_{n\to\infty} \frac{C(\pi)}{n(n-1)} = \langle d\rangle\langle t\rangle. \tag{6.13}$$

By using eq. 4.3 for the expected cost value in equilibrium, Bonomi and Lutton [BON86] derive eq. 6.13 in a straightforward way.

For the QAP we have (see also section 4.2)

$$\langle C(\infty)\rangle = \lim_{c\to\infty}\langle C(c)\rangle = \lim_{c\to\infty}\sum_{\pi} C(\pi)\cdot q_\pi(c) =$$

$$\frac{(n-2)!}{n!}\sum_{i=1}^{n}\sum_{j=1}^{n} d_{ij}\sum_{k=1}^{n}\sum_{l=1}^{n} t_{kl}, \tag{6.14}$$

which yields, using the law of large numbers:

$$\lim_{n\to\infty}\left(\frac{\langle C(\infty)\rangle}{n(n-1)}\right) = \langle d\rangle\cdot\langle t\rangle. \tag{6.15}$$

For arbitrary c we have (cf. eq. 4.10):

$$\langle C(c)\rangle = -\frac{\partial \ln Q(c)}{\partial(c^{-1})}, \tag{6.16}$$

where $Q(c)$ is the partition function[1] defined by eq. 4.9. For the QAP, we have

$$\exp(-C_{opt}/c) < Q(c) \le n!\cdot\exp(-C_{opt}/c), \tag{6.17}$$

which yields, using Stirling's formula and dividing by $n(n-1)$,

$$\frac{\ln Q(c)}{n(n-1)} = -\frac{C_{opt}}{c\cdot n(n-1)} + O\left(\frac{\ln n}{n}\right) \tag{6.18}$$

and

$$\lim_{n\to\infty}\frac{\ln Q(c)}{n(n-1)} = -\frac{1}{c}\lim_{n\to\infty}\frac{C_{opt}}{n(n-1)}. \tag{6.19}$$

[1] Bernasconi [BER87] uses the partition function in a similar way to obtain Golay's conjecture [GOLA82] for the *maximum merit factor* of long binary sequences.

Using eq. 6.16, one finally obtains

$$\lim_{n\to\infty}\left(\frac{\langle C(c)\rangle}{n(n-1)}\right)=\lim_{n\to\infty}\frac{C_{opt}}{n(n-1)}. \tag{6.20}$$

Combining eqs. 6.15 and 6.20, it is concluded that the left-hand side of eq. 6.20 is independent of c and equal to $\langle d\rangle\langle t\rangle$. Since the probability distribution in equilibrium (the stationary distribution of the homogeneous Markov chain, see section 3.1) clearly favours lower cost values for lower values of c, it can be concluded that eq. 6.13 holds **for each** permutation π.
Application of the simulated annealing algorithm to the aforementioned class of QAPs (with the same cooling schedule as for TSP and a transition defined as the inversion of the locations of two sources) indeed shows that as n goes to infinity the observed average value of the cost function does not alter significantly over consecutive Markov chains. Moreover, for small values of n, the final cost value found by the algorithm is considerably lower than the asymptotic value, given by eq. 6.15; for $n = 20$ a savings of 10% is reported.

(3) the Minimum Weighted Matching Problem (MWMP)

In the n-point MWMP (n even), a distance matrix $D = (d_{ij})$, $i, j = 1, \ldots, n$ is given. A *perfect matching* is defined as a set of $\frac{n}{2}$ edges such that each of the n points is the endpoint of exactly one of the $\frac{n}{2}$ edges. For n points, there are $(n-1)\cdot(n-3)\ldots 3\cdot 1$ such matchings. Let E_n denote the set of all perfect matchings of n points and let $b_\pi(i)$ and $e_\pi(i)$ denote the endpoints of the i-th edge ($i = 1, 2, \ldots, \frac{n}{2}$) for a perfect matching $\pi \in E_n$. The cost of π is defined as

$$C(\pi)=\sum_{i=1}^{\frac{n}{2}} d_{b_\pi(i)e_\pi(i)}. \tag{6.21}$$

The MWMP is the problem of minimizing $C(\pi)$ over E_n. The n-points MWMP can be solved exactly in $\mathcal{O}(n^3)$ time ([PAP82]), approximation algorithms running in $\mathcal{O}(n)$ or $\mathcal{O}(n^2)$ are reported to find final solutions on the average 30 to 50% above the global minimum [IRI83]. Consider the class of problems where the n points are again uniformly distributed on the unit square, where d_{ij} is the Euclidean distance between points i and j. If $C_{opt}^{(D)}$ is the optimal matching for an instance of this class with distance matrix D, then Papadimitriou [PAP77] shows that

$$\lim_{n\to\infty} \frac{C_{opt}^{(D)}}{\sqrt{n}} = \xi, \qquad (6.22)$$

where ξ is an unknown constant (by solving a large number of problems in this class exactly, it is conjectured that $\xi \in [0.32, 0.33]$ [IRI83]). Thus, asymptotically this class of MWMPs behaves similarly to the class of TSPs mentioned before (see eq. 6.11).
Lutton and Bonomi [LUT86] apply the simulated annealing algorithm (with a similar cooling schedule as for TSP and a transition defined as a replacement of a pair of edges by another pair, see figure 6.2) to this class of MWMPs and show that, as n goes to infinity, the cost value of the final configuration indeed approaches the conjectured value of $\xi \cdot \sqrt{n}$ within a few percent (values for ξ between 0.32 and 0.34 are reported). Moreover, the running time of the algorithm is $\mathcal{O}(n)$ (the Markov chain length is proportional to n, but the time to implement a transition is independent of n, as opposed to TSP). Thus, in this case, the simulated annealing algorithm runs as fast as the fastest approximation algorithm known so far ([IRI83]), but finds 30 to 50% better solutions.

6.4 Performance on combinatorial optimization problems

Johnson *et al.* [JOHN87], Kirkpatrick [KIR84], Morgenstern and Shapiro [MORC86], Nahar *et al.* [NAH85], [NAH86] and Skiscim and Golden [SKI83], [GOLDE86] discuss the performance of simulated annealing with respect to a number of well-known combinatorial optimization problems. Unfortunately, all experiments are carried out with simple cooling schedules (see section 5.2). Though such a simple cooling schedule might suffice to obtain satisfactory solutions, as the results obtained for large randomly generated problems in the previous section illustrate, the performance of the algorithm can be considerably improved by using a more elaborate schedule, based on the proposals described in section 5.3 [AAR85b]. In our opinion the conclusions based on these computational studies are premature and

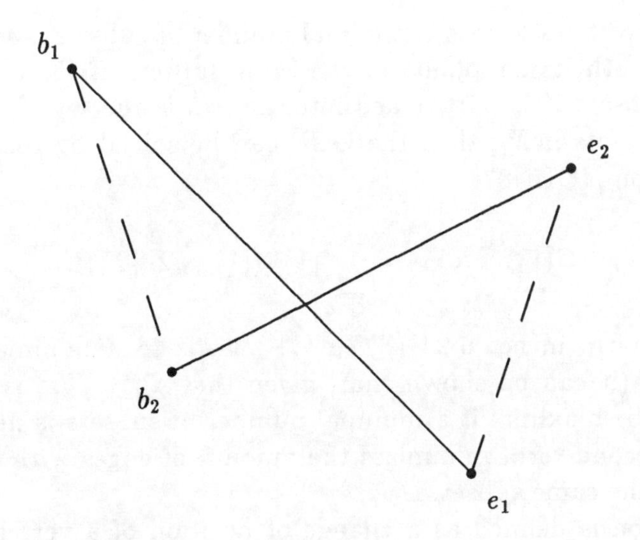

Figure 6.2: Example of a transition: edges (b_1, e_1) and (b_2, e_2) are replaced by (b_1, b_2) and (e_1, e_2).

of limited value.

Johnson *et al.* [JOHN87] discuss the performance of simulated annealing with respect to four problems: TSP (as described in the previous section), *graph partitioning* (GPP), *graph colouring* (GCP) and *number partitioning* (NPP).

In the GPP a graph $G = (V, E)$ is given and we are asked to partition the set of vertices V into two subsets V_1 and V_2 in such a way that the difference in cardinality between the two sets V_1 and V_2 and the number of edges with one endpoint in V_1 and one in V_2 are both minimal. If $\mathcal{E}$ denotes this set of edges with endpoints in different subsets, then the cost of a partition (V_1,V_2) is given by [JOHN87]

$$C(V_1, V_2) = |\,\mathcal{E}\,| + \lambda(|\,V_1\,| - |\,V_2\,|)^2, \tag{6.23}$$

where λ is a weighting factor. A transition is defined as a change of position of a vertex from the subset to which it currently belongs to the other subset.

In the GCP a graph $G = (V, E)$ is given and we are asked to partition the set of vertices V into a minimal number of subsets, such that no edge has both its endpoints in the same subset. If V_i, $i = 1,\ldots,k$ are the subsets V is partitioned into, and E_i is the set of edges with both endpoints in V_i, then the GCP can be solved by maximizing the function [JOHN87]

$$C(V_1,\ldots,V_k) = \sum_{i=1}^{k} |V_i| \, (|V_i| - 2 \cdot |E_i|). \tag{6.24}$$

The first term in eq. 6.24 ($\sum_{i=1}^{k} |V_i|^2$) leads to a minimal number of subsets (it can be shown that, given that $\sum_{i=1}^{k} |V_i|$ is constant, $\sum_{i=1}^{k} |V_i|^2$ is maximal if a minimal number of subsets is non-empty) and the second term minimizes the amount of edges with both endpoints in the same subset.
A transition is defined as a change of position of a vertex from its present subset to another subset.

Finally, in the NPP, n rational numbers $r_1,\ldots,r_n$ in $(0,1) \subset \mathbf{Q}$ are given and we are asked to partition them into two sets Q_1 and Q_2 such that the cost function, given by

$$C(Q_1,Q_2) = \left| \sum_{i \in Q_1} r_i - \sum_{j \in Q_2} r_j \right|, \tag{6.25}$$

is minimized. A transition is defined by changing the position of a randomly chosen rational number from its current subset to the other one.

TSP, GPP, GCP and NPP are all NP-hard ([KAR72], [GAR76]) and good polynomial-time approximation algorithms are therefore of prime interest. Johnson *et al.* [JOHN87] investigate whether or not simulated annealing can be viewed as such an algorithm. The following cooling schedule is used:

1. initial value of the control parameter given by eq. 5.7;
2. decrement rule for c given by eq. 5.8;
3. length of Markov chains determined by a minimal number of accepted transitions, bounded by a multiple of the size of the neighbourhoods;

4. the stop criterion is satisfied when no improvements have been found during a fixed number of consecutive Markov chains.

Johnson *et al.* apply simulated annealing with this cooling schedule to a large number of (randomly generated) instances of the aforementioned four problems. From the comparisons between their results and those obtained by other approximation algorithms, the following conclusions are reached:

1. for the GPP, simulated annealing obtains final solutions that are at best some 5 percent better than those obtained by the best of the more traditional algorithms ([KER70], [FIC82] and [GOLDB83]) if the latter are allowed the same amount of computation time as simulated annealing. For sparse graphs, the difference between solutions obtained by simulated annealing and by for example repeated application of Kernighan-Lin heuristic [KER70] can be as much as 70% in favour of simulated annealing (reported for 1000 vertices, average number of edges/vertex = 2.5). For some structured graphs, however, the Kernighan-Lin algorithm finds better solutions.

2. solving NPPs by simulated annealing results in final solutions deviating substantially from a global minimum as is also the case for other approximation algorithms [JOHN87]. This can be explained by the fact that if the rational numbers are independently and uniformly distributed on the interval $(0,1)$, the expected minimum cost value is $\mathcal{O}(\frac{\sqrt{n}}{2^n})$ [JOHN87], whereas the smallest change in the cost function that can be made by one transition is expected to be $\mathcal{O}(\frac{1}{n})$. Since, for $n \gg 1$, $\frac{\sqrt{n}}{2^n} \ll \frac{1}{n}$ the differences in cost are too coarse to give any guidance for the algorithm to a global minimum. When transitions consisting of a change in position of k rational numbers are used, the result is an expected difference in cost of $\mathcal{O}(\frac{1}{n^k})$. Thus, when k is sufficiently large, the expected difference in cost is of the same magnitude as the minimum cost value. Unfortunately, Johnson *et al.* do not investigate whether this type of transition yields better solutions.

3. for the GCP, simulated annealing produces final configurations that are competitive with those generated by a tailored heuristic (the one of Johri and Matula [JOHR82]). This heuristic is known as the best one for this problem. However, computation times for simulated annealing are considerably longer than those of the specialized heuristic (for a 1000-vertex graph, computation times of up to 22 hours are reported).

4. for the TSP, simulated annealing consistently outperforms solutions found by repeated application of iterative improvement, based on 2-opt or 3-opt transitions, but it is a consistent loser when compared with the well-known Lin-Kernighan algorithm [LIN73], which is based on k-opt transitions, where at each iteration the algorithm decides dynamically what the value of k should be ($k \geq 2$).

In an earlier paper, Kirkpatrick [KIR84] also studied the performance of simulated annealing when applied to the GPP and the TSP. His cooling schedule is very similar to the one used by Johnson *et al.* [JOHN87]. For the GPP, Kirkpatrick compares simulated annealing with Goldberg and Burstein's [GOLDB83] implementation of the Kernighan-Lin heuristic [KER70] and reaches the same conclusions as Johnson *et al.* but for the TSP his conclusions are based on comparisons of simulated annealing with iterative improvement techniques based on 2-opt or 3-opt transitions. Like Johnson *et al.* Kirkpatrick concludes that simulated annealing compares favourably with these heuristics, but there is no attempt to compare simulated annealing with Lin-Kernighan's k-opt heuristic.

In a recent paper, Morgenstern and Shapiro [MORC86] also report on the performance of simulated annealing applied to the GCP, but they use a transition mechanism differing from the one employed by Johnson *et al.* in that only *legal* partitions are generated, i.e. partitions of the set of vertices V for which all sets E_i are empty. The latter is realized by using *Kempe chains*: if V_i and V_j are two subsets of the partition and v is a vertex in V_i, then the $i-j$ Kempe chain containing v consists of the connected subgraph of G obtained by searching outward from v, traversing only edges connecting vertices in either V_i or V_j. The change of a vertex in V_i to another subset V_j does not effect the legality of the partition if **all** vertices in the $i-j$ Kempe chain are also changed (from V_i to V_j or vice versa). Thus, the transition mechanism is implemented as follows. Given that the number of subsets in the present partition is K, choose i and j at random from $[1, K]$ and $[1, K+1]$, respectively, choose a random vertex v in V_i and then change the position of all vertices in the $i-j$ Kempe chain containing v (note that if $|V_i| = 1$, the cost function decreases by 1, whereas it increases by 1 if $j = K+1$). Using this transition mechanism, the algorithm needs significantly less computation time than the implementation of Johnson *et al.* to obtain the same quality of results (for the 1000-vertex graph, for example, the computation time reduces from 22 to 4.5 hours on a VAX 11/780-computer, the final partitions contain 99 and 93 subsets, respectively). These computation times are competitive with those of the specialized heuristic ([JOHR82]).

Another computational study in which simulated annealing is compared with other algorithms is reported by Nahar *et al.* [NAH85]. The two problems under consideration are the *Net Linear Arrangement Problem* (NLAP) and the *Graph Linear Arrangement Problem* (GLAP). In the NLAP, n nodes and m nets are given, each net connecting a number of nodes, and we are asked to obtain a linear ordering of the nodes such that the number of net crossings between any pair of adjacent nodes is as small as possible. The special case of the NLAP, in which each net connects exactly two nodes, is called GLAP (in subsection 7.2.2 it is shown that this problem can be solved by repeatedly solving the GPP).
Nahar *et al.* use the following cooling schedule:

1. initial value of the control parameter given by 1;

2. a fixed set of values of the control parameter (containing 1,2 or 6 elements);

3. length of Markov chains determined by a minimal number of **rejected** transitions (!).

The acceptance matrix is introduced as an extra parameter, $A_{ij}(c)$ can be chosen as a linear, quadratic, cubic or exponential function in ΔC_{ij} or even piecewise constant. In the latter case, setting

$$A_{ij}(c) = \begin{cases} \varpi & \text{if } \Delta C_{ij} > 0 \\ 0 & \text{if } \Delta C_{ij} \leq 0, \end{cases} \tag{6.26}$$

simply implies that any deterioration is accepted with probability ϖ. For the GLAP, experiments are carried out on a large set of randomly generated graphs and it is concluded that there is no difference in quality of final solution obtained by simulated annealing (with six values of c and $A(c)$ given by eq. 3.17) and a schedule in which all transitions are accepted ($A_{ij}(c) = 1, \forall i, j$). Since the latter simply implies a random walk through the set of configurations, this schedule is implemented such that merely each κ-th deterioration is accepted with probability 1, all other deteriorations are rejected (in the experiments κ is set to 18). Similar results are found for the NLAP. Since the second schedule does not involve any choice of values for c, the authors conclude that it is to be preferred to simulated annealing (with its inherent difficulties of parameter choice). Because of the extreme simplicity and possible incorrectness (see 3. above) of the employed cooling schedule, we must strongly doubt this conclusion. Moreover, the conclusions are more or less contradicted by those of Johnson *et al.* [JOHN87], since the last authors claim the related GPP to be a success story for simulated annealing. A sequel to this paper [NAH86], in which similar results are reported, suffers from similar methodological flaws, e.g. the condition for terminating execution of the algorithm is given by "continue until out of time", so that for a given small amount of computation time (as in the experiments carried out by the authors) a final solution, which is not even locally optimal with respect to the underlying neighbourhood structure, is to be expected. Moreover, the authors do not seem to realize that the performance of simulated annealing is independent of the initial configuration, witness the fact that they investigate the effect of the initial configuration on the quality of the final solution.

Skiscim and Golden [SKI83], [GOLDE86] study the performance of simulated annealing on the TSP. Randomly generated TSPs as well as TSPs from the literature are considered. The cooling schedule used by Skiscim and Golden, was already partly described in section 5.2 (length of Markov chains), furthermore, a fixed set of values of the control parameter is used. The conclusions of Skiscim and Golden are roughly the same as those of Johnson *et al.* for TSP: simulated annealing is consistently outperformed, both in terms of quality of final solution and computation time, by a specialized heuristic for TSP (in this case the CCAO-procedure [GOLDE85]). We repeat our objection against such conclusions, based on applying simulated annealing with a conceptually simple cooling schedule.

In [RAN86], Randelman and Grest present a systematic empirical analysis of the performance of the simulated annealing algorithm on randomly generated instances of the TSP. Applying a cooling schedule with a fixed difference Δc ($\Delta c > 0$) between successive values of the control parameter and a fixed Markov chain length L, they conclude that the expected final cost $\langle C_{fin} \rangle$ obtained by the algorithm depends logarithmically on the cooling rate $\frac{\Delta c}{L}$, i.e.

$$\langle C_{fin} \rangle \propto -\left(\ln \frac{\Delta c}{L}\right)^{-1} . \tag{6.27}$$

In [AAR88a], Aarts *et al.* report on a systematic pseudo-empirical analysis of the average running time and quality of the final solution obtained by the simulated annealing algorithm for a special class of TSPs applying the cooling schedule of Aarts and Van Laarhoven (see section 5.3). From this analysis it is concluded that the probability to obtain a final solution deviating less than ε ($\varepsilon > 0$) from the minimum tour length is given by

$$Pr\{C_{fin} - C_{opt} < \varepsilon \mid \delta\} = \frac{1}{\Gamma(p)} \sum_{n=0}^{\infty} \frac{(-1)^n \varepsilon^{p+n}}{(p+n)n!}, \tag{6.28}$$

where

$$p = a[\ln(1+\delta)]^b; \tag{6.29}$$

δ is the parameter determining the cooling rate (see eq. 5.32) and a and b are parameters depending on the problem instance. Furthermore, it is concluded that the corresponding running time depends exponentially on the factor $\ln(1+\delta)$. Combination of these two results leads to a similar observation as reported by Randelman and

Grest [RAN86], i.e. the quality of the final solution depends logarithmically on the cooling rate.

The effect of using a more elaborate schedule is illustrated by Aarts and van Laarhoven [AAR85a], [AAR85b], [AAR89], [LAA88b]. In [AAR85a], the performance of simulated annealing on a special class of TSPs is studied. The TSPs under consideration are those where the locations of the cities are positioned on a regular grid, see figure 4.1. As a consequence, the minimum length of a tour is known **a priori**, which is a useful tool in the analysis of the performance of the algorithm. The employed cooling schedule is the one discussed in section 5.3 (see Table 5.1). Aarts and Van Laarhoven show that, by choosing δ (determining the decrement in c, given by eq. 5.32) sufficiently small, the minimum cost value can be approached within any desired margin [AAR85a], [AAR89], [LAA88b]. Evidently, smaller values of δ result in larger running times, but according to eq. 6.4, even for small values of δ the running time is still polynomial in the size of the problem (in this case, we have, since $R = \frac{1}{2}n(n-1)$, $|\mathcal{R}| = \frac{1}{2}(n-1)!$ and each transition takes $\mathcal{O}(n)$ to implement, an $\mathcal{O}(n^4 \ln n)$ running time). Only asymptotically (for $\delta \downarrow 0$), the running time is exponential in the size of the problem.

Aarts and Van Laarhoven [AAR89] apply the same schedule to a number of well-known TSPs from the literature. In Table 6.1, the final deviation from the minimum tour length obtained by simulated annealing for five of these problems as well as the computation times (in minutes on a VAX 11/785-computer) are given. The results are averaged over five runs of the simulated annealing algorithm (with different seeds for the random number generator) with the aforementioned cooling schedule and $\delta = 0.1$. As can be concluded from these results, it is indeed possible to obtain tours close to an optimal one at the expense of relatively large amounts of computation time. The 100-city problem of Krolak *et al.* [KRO71] and the 318-city problem of Lin and Kernighan [LIN73] are also part of the computational study of Golden and Skiscim [GOLDE86]. For the 100-city problem the best solution obtained by their implementation of simulated annealing is 1.34% above the global minimum (found in 22.6 minutes on a VAX 11/780-computer), for the 318-city problem the shortest tour length

Problem	*# cities*	C_{opt}	$\overline{C}_{final}$	$\overline{\Delta}$ *in %*	*CPU*
[GRÖ77]	48	5046	5094.8	0.97	1.6
[THO64]	57	12955	13068.0	0.87	2.6
[KRO71]	100	21282	21467.8	0.87	13.7
[GRÖ77]	120	6942	7057.2	1.66	22.8
[LIN73]	318	41345	41957.4	1.59	399.0
[GRÖ84]	442	5069	5147.0	1.54	978.9

Table 6.1: Average performance of simulated annealing on 5 well-known TSPs, C_{opt} denotes minimum tour length, $\overline{C}_{final}$ average final tour length after annealing, $\overline{\Delta}$ average deviation of final cost from C_{opt} in percentages, CPU times in minutes on a VAX 11/785-computer (reproduced from [AAR89]).

found by annealing is 4.03% above the minimum tour length (found in 713.6 minutes). The large difference between these numbers and the **average** results shown in table 6.1 is another indication for the importance of choosing a more elaborate cooling schedule.
Lam and Delosme [LAM86a] also compare a simple schedule with a schedule, whose decrement rule is similar to eq. 5.32, by solving TSP's and claim a hundredfold speedup of the latter schedule over the former one with the same quality of solution.
Aarts and Van Laarhoven [AAR85b] consider the effect of a cooling schedule on the performance of the simulated annealing algorithm by solving a large set of randomly generated GPPs. Three cooling schedules are compared (see Table 5.1):

1. the schedule of Aarts and Van Laarhoven [AAR85a];

2. a schedule based on the proposals of Romeo and Sangiovanni-Vincentelli [ROM85];

3. a conceptually simple cooling schedule.

The parameters in schedules 2 and 3 are determined empirically. From the results it is concluded that schedule 1 yields an average improvement of 13.1% over the results found by iterative improvement, as compared with 6.8% and 10.7% for schedules 2 and 3, respectively. Moreover, since all experiments are carried out on the same computer and with the same computer code, computation times for the different schedules can be properly compared. It is concluded that schedule 1 has the smallest computation times.

Rossier *et al.* [ROSS86] also study the TSP. With a decrement rule given by

$$c_k = \gamma_i, \; k_i \leq k < k_{i+1}, \tag{6.30}$$

$$\gamma_i = \left(\frac{1}{\tau}\right)^i, \; i = 1, 2, \ldots, \tag{6.31}$$

$$k_i = k_{i-1} + K, \; i = 1, 2, \ldots, \tag{6.32}$$

for some constants τ and K and a Markov chain length proportional to the number of cities, a solution to Grötschel's 442-city problem [GRÖ84] with tour length 51.42 inches is found in 238 seconds on a Cyber 170-855-computer. The optimum tour length for this problem is 50.69 inches [GRÖ87]. As in Bonomi and Lutton's work on TSP [BON84], transitions are restricted to cities belonging to the same or adjacent subregions into which the set of all cities is partitioned.

Finally, Burkard and Rendl [BUR84] study the QAP (cf. section 6.3). Using a cooling schedule of the type discussed in section 5.2 ($\alpha = 0.5$, Markov chain length proportional to the number of variables n), results are obtained for a number of problems from the literature that are comparable with those found by a tailored heuristic for this problem (the one described in [BUR83b]), but in far less computation time.

Preluding on chapter 9, we draw the following preliminary conclusions:

- the performance of the simulated annealing algorithm depends strongly on the chosen cooling schedule; this is especially true for the quality of the solution obtained by the algorithm;
- with a properly chosen cooling schedule near-optimal solutions can be obtained for many combinatorial optimization problems;
- computation times can be long for some problems.

Chapter 7

Applications

7.1 Introduction

Ever since its introduction in 1982, simulated annealing has been applied to many combinatorial optimization problems in such diverse areas as *computer-aided design of integrated circuits*, *image processing*, *code design*, *neural network theory*, etc. In many of these areas, previous algorithms, if existing at all, performed quite poorly and the territory was ripe for an easy-to-implement approach with a modicum of sophistication [JOHN87].
In this chapter the applications of simulated annealing in one of these problem areas, computer-aided design of electronic circuits, are extensively discussed (section 7.2); applications in other areas are briefly summarized in section 7.3, where the appropriate references are given. We consider the combinatorial optimization problems in computer-aided design to be representative for the kind of problems simulated annealing can be successfully applied to: large-scale optimization problems, for which simple optimization strategies as for example iterative improvement (see chapter 1) fail to find good approximations to an optimal solution. Furthermore, computer-aided circuit design is probably the area to which so far the majority of applications of simulated annealing belong. Finally, it is the area the present authors are most familiar with.
The performance of simulated annealing in more competitive areas, where good approximation algorithms are already in existence (TSP,

graph partitioning), was discussed in the previous chapter; in the applications discussed in this chapter, the emphasis is above all on solving the problem rather than on the performance of the algorithm. The material is chosen such that it brings out a number of different and interesting aspects related to the application of simulated annealing to a variety of optimization problems, illustrating the general applicability of the algorithm. Moreover, many of these problems belong to the class of NP-hard problems (see chapter 1).
Application of the simulated annealing algorithm to a combinatorial optimization problem presupposes the definition of:

- a way to represent configurations,
- a cost function, and
- a generation mechanism for transitions.

For each of the applications the discussion is gathered around these definitions. Where possible, the employed cooling schedule is discussed as well.

7.2 Applications in computer-aided circuit design

7.2.1 Introduction

The design and development of cost-efficient and reliable electronic systems necessitates the use of integrated circuits, especially high-density, high-performance, very large scale integrated (*VLSI*) circuits [CART86]. As the complexity of VLSI-systems increases, the use of computer-aided design techniques is inevitable to shorten design times, to minimize their physical size and to ensure their reliability. To accomplish this task effective techniques must be established to automate various steps in the design of VLSI chips (*design automation*). For a thorough introduction to the state of the art of design automation for integrated circuits, the reader is referred to the April 1986-issue of *Computer* [COM86], which is devoted solely to this subject.

The objectives of design automation are realized by relieving design engineers from time-consuming and repetitive tasks such as for example the painstaking optimization problems that arise during the design of an integrated circuit. Many of these optimization problems are of a combinatorial nature, which explains why simulated annealing is widely applied as an optimization algorithm in this area. Many of these problems are *layout problems*: placing box-like modules in a plane and subsequently connecting them by wires in such a way that the area occupied by modules and wires is as small as possible [BRE72], [SOU81]. These *placement* and *routing* problems are discussed in subsections 7.2.2 and 7.2.3, respectively. Other applications to combinatorial optimization problems in computer-aided circuit design are discussed in subsection 7.2.4.

7.2.2 Placement

The placement problem is a problem simulated annealing has been particularly successfully applied to, as witness the many papers devoted to this subject as well as a number of papers on parallel implementations of placement algorithms by simulated annealing (see section 8.1). Within IBM, for example, simulated annealing is now widely accepted and used to solve several kinds of placement problems [ROT86b].

In the most general case of a placement problem, a set of rectangular *modules* (or *cells*) of variable size is given as well as connectivity information: a list of *nets*, each net specifying a set of modules that are to be connected. The problem is to place the modules in an area of minimal size and such that the subsequent *routing*, the positioning of the connecting wires between the modules, takes as little area as possible ('placement with routing in mind'). Thus, the cost function must incorporate, for each placement, the total area occupied by the modules (e.g. the area of the *bounding rectangle*) as well as an estimation of the cost of the subsequent routing (obviously, it is undesirable to actually solve a routing problem for each placement).

The general case, with modules of different size, is considered by Jepsen and Gelatt in [JEP83] and by Sechen and Sangiovanni-Vincentelli in [SEC85], [SEC86b]. Besides, the positions in which a module can be placed are not restricted, e.g. to a one-dimensional array or to a regular grid. **A configuration** is defined as a set of locations of all modules (a placement). Moreover, to avoid complications in defining transitions, modules can be located such that they (partly) overlap each other. By allowing overlap, the connectivity of the set of configurations is enlarged. However, configurations containing overlap are illegal (physically not realizable) and therefore the overlap is usually incorporated in the cost function as a *penalty function.* In this way, the algorithm strives for configurations with a minimal amount of overlap.

In [JEP83], the **cost function** consists of two components. The first one estimates the quality of solution of the (subsequent) routing problem by defining a number of (imaginary) horizontal and vertical lines across the bounding rectangle and counting the number of nets crossings on each line. For each line, this number is squared and the sum of all these squares is the first cost component. The second component represents the total amount of overlap in the present configuration.
The **transitions** in [JEP83] consist of random length translations of modules in horizontal or vertical directions, 90 degree rotations in either direction and vertical or horizontal reflections of modules (excluding overlap would necessitate checking the legality of these transitions). In [JEP83], an example is presented consisting of several tens of modules, but without any reference to implementation details (e.g. the cooling schedule employed).

The *Timber Wolf* package, developed by Sechen and Sangiovanni-Vincentelli [SEC85], [SEC86a], [SEC86b] (a preliminary version is described in [SEC84]), is an integrated set of placement and routing optimization algorithms, all of which use simulated annealing as optimization technique. The cooling schedules used in TimberWolf's algorithms are of the type discussed in section 5.2 (Markov chain length proportional to the number of modules, fixed decrement rule for the control parameter - in a more recent version [SEC86a] a slightly different choice is made).

One of the algorithms, the *macro/custom cell* placement algorithm can be used to solve the aforementioned placement of modules of different size. However, there are two additional degrees of freedom. Firstly, the modules can be such that they are known only to occupy an estimated area, i.e. the product of width and length is fixed. In that case, the algorithm determines an optimal *aspect ratio* (ratio of width and height of a module) for such a module within given minimum and maximum values. Secondly, *pins* (locations where nets connect to a module) do not have to be specified, it suffices to specify which particular side of a module they are to be assigned to. The algorithm then determines an optimal location for each pin on the specified side.

A **configuration** is given by a set of locations of all modules and (possibly) by sets of pin locations and aspect ratios. The **cost** of a configuration is composed of two parts: the total estimated wire length of connections (based on the sum over all nets of the perimeter of a net's *bounding box*, defined as the smallest rectangle enclosing all pins comprising a net) and a *penalty* term, measuring the degree to which a configuration falls short of requirements. The latter can occur in two ways: when modules overlap each other or when the number of pins on a particular location on the border of a module exceeds the capacity of that location. Furthermore, the penalty term is defined such that as the control parameter approaches zero, overlap is increasingly penalized, so that there is a minimum amount of overlap (possibly none) in the final configuration.

Transitions are similar to those in [JEP83], but due to the additional degrees of freedom, it is also possible to change a pin location or to change the aspect ratio of a module. Furthermore, the displacement of a module to a new location is controlled by a *range limiter*, which limits the range of such a displacement in the final stages of the algorithm (cf. the concept of transition classes introduced by White [WHI84] and discussed in section 5.4).

Results are reported for a 613-modules/900-nets problem, each module having 2 to 64 pins. In 18 hours of CPU time on a VAX 11/780-computer a solution is found whose estimated length of connections is 21% less than that of a manual placement (which required four months of effort), clearly a striking result.

The placement of arbitrarily sized rectangular modules is also consid-

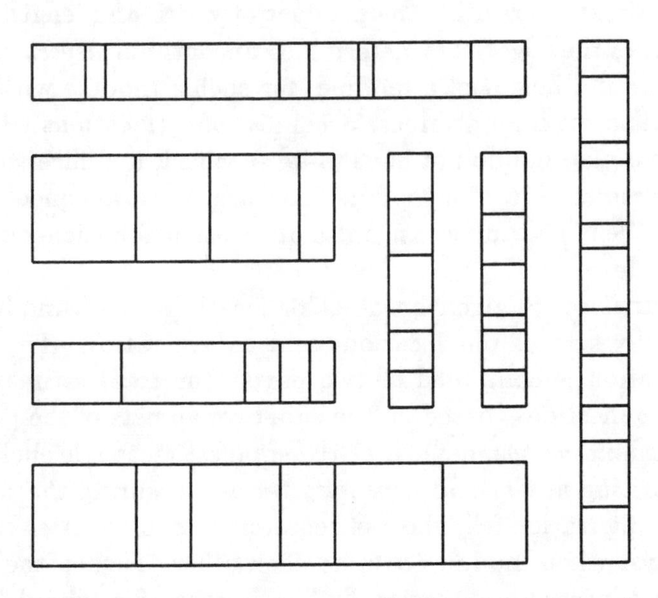

Figure 7.1: Example of standard cell placement (after [SEC85]).

ered in an introductory paper on simulated annealing by Siarry *et al.* [SIA85]. In this paper, the **cost** of a placement is defined as the sum of total estimated wire length and the product of the initial total estimated wire length and the total amount of overlap. The latter term is multiplied by a constant factor, which gradually increases from 0 to 1 as the control parameter approaches 0, so as to increase the penalty for overlap for a given configuration.

Another algorithm in TimberWolf solves *standard cell* placement problems (an extensive description of this part of TimberWolf can be found in [SEC86a], the complete standard cell layout system of Berkeley is described by Braun *et al.* [BRAU86]). In this type of problem, all modules are to be placed in horizontal or vertical rows (called *blocks*) and each horizontal (vertical) block contains modules of the same height (width) but varying width (height), see figure 7.1. Thus, there are restrictions with respect to both the sizes

of the modules and their locations.
A configuration is given by a set of locations of all modules in the blocks. The **cost function** is similar to the one in the macro/custom cell placement algorithm and consists of the total estimated length of connections and a penalty term for overlap as well as the amount by which the sum of the lengths of the modules in a particular block exceeds the actual length of a block.
Transitions are defined as pairwise interchanges of modules (possibly blocks) or displacements of modules to a new location. Both types of transitions are again controlled by a range limiter.
The algorithm is compared with the CIPAR standard cell placement package developed by American Microsystems Inc. For the larger problems (800 to 1700 modules) the total estimated length of connections found by TimberWolf is 45 to 66 percent less than the one found by CIPAR, again a striking result. The largest problem (1700 modules) takes 84 hours of CPU time on a VAX 11/780-computer.
Similar results obtained in significantly less computation time are reported by Huang *et al.* [HUA86], where the performance of TimberWolf with the cooling schedule of Huang *et al.* (see section 5.3) is evaluated. For a 469-cells circuit, for example, the CPU time on a VAX11/780-computer reduces from more than 3 hours to less than 1.5 hours.
In [GRO86], Grover shows how the probabilistic nature of the simulated annealing algorithm can be exploited to reduce computation times, by carrying out cost calculations *approximately* instead of exactly and illustrates this by means of standard cell placement. Given that the probability to accept a deterioration in cost is given by $\exp\left(\frac{-\Delta C}{c}\right)$, we can expect this probability not to change significantly if ΔC is replaced by $\Delta C \pm \delta$, where $\delta \ll c$. In the standard cell placement algorithm, such replacements occur when only occasionally updates of the positions of all cells take place, so that most of the time the algorithm carries out its cost calculations not knowing the exact positions of all cells. It is shown that this leads to a 3 to 5 times faster algorithm for results of the same quality.

The third placement algorithm in TimberWolf solves *gate-array* placement problems. In this type of problem, all modules have the same size and are to be positioned on a regular grid, see figure 7.2. The

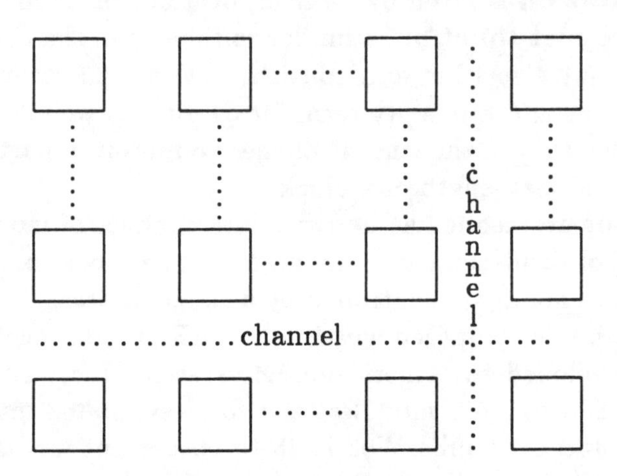

Figure 7.2: Example of gate-array placement.

horizontal and vertical strips between rows and columns of modules are referred to as *channels.* The objective is to position the modules in such a way that the channels, containing the connections, can be as small as possible.
A **configuration** is given by an assignment of all modules to grid points. The **cost function** is based on *net crossing histograms* for the vertical and horizontal channels (first introduced by Kirkpatrick *et al.* [KIR82]). Such a histogram is computed by considering the bounding box of each net and adding 1 to the histogram for each channel intersecting the bounding box. **Transitions** are defined as pairwise interchanges of modules; because all modules have the same size overlap cannot occur.
The performance of TimberWolf is discussed by solving problems reported by Stevens [STE72]. For the largest example (151 modules to be placed on a 11 $\times$ 15 grid) TimberWolf reduces the total estimated wire length by 21% compared with Steven's result and by 17% over the result reported by Goto and Kuh [GOT78]. Total computation time is 15 minutes on a VAX 11/780-computer.

A similar approach is reported by Jess *et al.* [JES84], where a gate-array design system is described which uses simulated annealing to solve the corresponding placement problem, and in introductory papers on simulated annealing by Siarry *et al.* [SIA84] and [SIA85]. Storer *et al.* [STO85] use simulated annealing to solve a gate-array-like placement problem: placement of modules of approximately the same size. The **cost function** is based on what are referred to as *routing functions* (it is beyond the scope of this monograph to explain this notion in more detail) and it is claimed that these functions probably yield better estimates of total wire length than the aforementioned histograms. However, no computational results are given to support this claim.

An even more restricted form of placement is the Net Linear Arrangement Problem (NLAP, see also section 6.4), where the problem is to position a number of modules in a linear array in such a way that the connections can subsequently be routed in as small an area as possible. Two approaches to solve this problem are presented by Aarts *et al.* [AAR84], a direct approach and a recursive approach. In both approaches, a **configuration** is given by a permutation of the set $\{1, 2, \ldots, n\}$, representing a sequence of the n modules. In the direct approach, the NLAP is solved by minimizing a **cost function** consisting of a term representing the maximum number of net crossings between two adjacent modules and a second term equal to the total number of net crossings in the array. The second term is added to favour configurations that have fewer net crossings from among all configurations with the same maximum number of net crossings. **Transitions** are again defined as pairwise interchanges of modules.
In the recursive approach, the NLAP is solved by recursively solving Net Partitioning Problems (NePPs). The NePP is a generalization of the GPP (which was already described in section 6.4) to multi-module nets, i.e. nets connecting more than two modules. The **cost function** in an NePP is similar to the one in a GPP (see eq. 6.23) and so are **transitions**. An NLAP can now be solved by first partitioning the array into two sets (an NePP), followed by partitioning each of these two sets, and so on, until the sets finally consist of single modules.
The largest example in [AAR84] is a 17-modules/45-nets problem, for

Example	*Gates (size)*	*Annealing*		*Iter. impr.*		*Min-cut*	
		WL	*t*	*WL*	*t*	*WL*	*t*
1	14	214	13	214	17	210	2
2	21	221	15	222	22	419	2
3	26	411	51	411	66	525	4
4	33	469	41	470	56	605	4
5	44	1075	223	1084	304	1313	9
6	51	1171	308	1175	388	1690	11
7	57	1290	587	1292	685	1994	17
8	72	1497	322	1500	369	2046	120
9	253	8047	8640	9433	10560	7851	103
10	305	19705	7980	20582	9600	20577	105

Table 7.1: Estimated wire length (WL) and computation time t (in seconds on a VAX 11/780) for different placement algorithms (reproduced from [ROW85b]).

which a placement is found in 53 seconds on a VAX 11/780-computer (following the direct approach). Numerical experiments indicate that the direct approach performs slightly better than the recursive approach [AAR84].

Solving NLAPs via the recursive approach is also reported by Rowen and Hennessy [ROW85a], [ROW85b], where NLAPs occur while generating automatically a special type of circuit called *Weinberger Array's* [WEI67]. Simulated annealing is compared with a few other algorithms, including iterative improvement and *min-cut linear arrangement*, a recursive approach using the Kernighan-Lin heuristic [KER70] to solve the NePPs. In most cases, simulated annealing obtains the smallest value for the estimated wire length but in the largest computation time; see Table 7.1.

The placement problem for *gate-matrix layouts* [LEO86], [WOU86] is a generalization of the NLAP because of the fact that for some nets,

the set of modules[1] connected by a net is not fixed, thus introducing an extra degree of freedom. In [WOU86], Wouters *et al.* take this degree of freedom implicitly into account by introducing dummy modules. Suppose, for example, that in an instance of this problem net i connects modules a, b and c, whereas net j connects either modules a, d and e or modules c, d and e. In this case, a dummy module m is introduced and an NLAP is solved, in which net i connects modules m, a, b and c and net j connects modules m, d and e. In the final solution the dummy modules are removed to obtain a solution to the gate-matrix placement problem.
Leong [LEO86] takes the extra degree of freedom explicitly into account: each **configuration** is not only determined by a permutation of the modules but additionally by a fixed set of modules for each net. **Transitions** are defined as a pairwise interchange of two modules or a change of the set of modules connected by a net. The problem is then solved along the same lines as solving an NLAP via the direct approach. For a problem consisting of 71 modules and 131 nets, the maximum number of net crossings in the final solution found by Leong is 29; in previous approaches to this problem, where it was impossible to take the extra degree of freedom into account, the best solution reported for the 71-module problem had a maximum number of net crossings equal to 40.

The placement problems for Weinberger arrays and gate-matrix layouts are also addressed by Devadas and Newton in [DEV86a], where a general array-placement program based on simulated annealing is described.

To conclude we mention that Dunlop and Kernighan [DUN85] solve standard-cell placement problems by using a graph partitioning approach. The graph partitioning problems are solved by using the Kernighan-Lin heuristic [KER70], but the authors devote a few words to an approach via simulated annealing and conclude that simulated annealing finds partitions of the same quality as the Kernighan-Lin heuristic, when allowed the same amount of computation time.

Summarizing, we conclude that simulated annealing can be success-

[1] In gate-matrix layouts modules are called *signals*.

fully applied to a large class of placement problems, sometimes at the expense of long computation times. Furthermore, in some cases the success of the algorithm depends to a great extent on the skill of the user in choosing an appropriate cost function.

7.2.3 Routing

In the design of VLSI circuits, placement is usually followed by routing, the positioning of the wires or nets between the modules. In this subsection we discuss two routing problems for which simulated annealing has been proposed as an optimization technique, the *global routing* problem (Vecchi and Kirkpatrick [VEC83], Sechen and Sangiovanni-Vincentelli [SEC85]) and the *channel routing* problem (Leong *et al.* [LEO85a], [LEO85b]).

Once the positions of the modules are known, it is customary first to construct a global routing for each wire, i.e. without specifying the exact location of the wires. The global router decides through which channels individual connections will run without specifying the exact locations in the channels. For the global routing problem considered by Vecchi and Kirkpatrick [VEC83] it is assumed that the modules can be thought of as being positioned on a regular grid (see figure 7.3). Furthermore, it is assumed that each net connecting more than two modules is represented by a set of nets between **pairs** of modules (a net connecting k modules, $k \geq 2$ can be broken into $k-1$ nets between pairs of modules). The global routing problem now consists of determining the links (the connections between adjacent grid points) each net is to occupy in such a way that the distribution of the nets over the links is as uniform as possible, the latter to reduce the likelihood of overflows. An overflow arises when the number of nets which is to be routed through a link exceeds the capacity of that link. The solution to the global routing problem can be used as a starting point for the final routing phase, in which the exact locations of the individual nets in the channels are determined.
If m_ν is the number of nets on link ν, the objective is to minimize

$$F = \lambda \sum_{\nu} {m_\nu}^2. \tag{7.1}$$

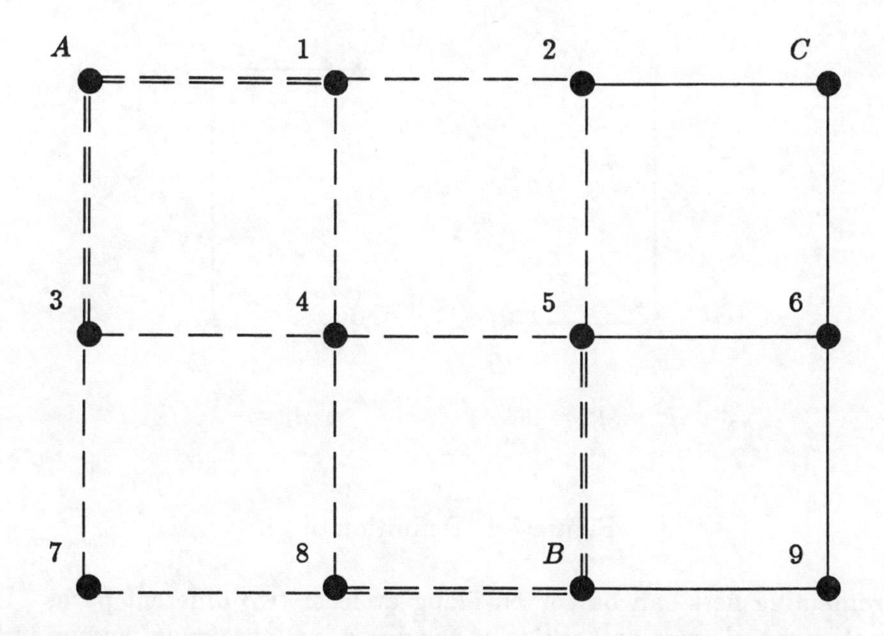

Figure 7.3: Model of the placement of modules, represented by dots. The broken lines (*A*-3-7-8-*B*, *A*-3-4-5-*B*, *A*-1-4-8-*B* and *A*-1-2-5-*B*) denote the 4 possible ways in which net *AB* can be routed (after [VEC83]).

To explain why the squared values are used, consider an increase of m_ν by 1, which yields an increase of F by $2 \cdot m_\nu + 1$. Thus, the larger the value of F, the greater the penalty for increasing F.
Traditionally, such routing problems are solved by routing the nets **one after the other**, which poses the difficult problem of determining the best order in which the nets are to be routed. In the simulated annealing approach, however, all nets are considered **simultaneously** (the objective function has a component for each net). Clearly, this allows more degrees of freedom.
To derive an expression for m_ν, assumptions must be made as to how nets are to be routed. When the two modules lie on a straight line, it is assumed that the net is routed via the minimum length path (in figure 7.3, modules *A* and *C* would be connected via *A*-1-2-*C*). The

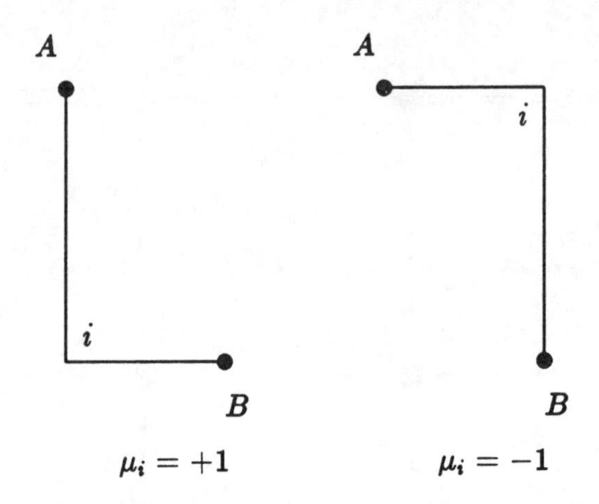

Figure 7.4: Definition of μ_i.

remaining nets can be routed along at least two different paths. If paths are allowed to have only one bend, we obtain, in Vecchi and Kirkpatrick's terminology, *L-shaped* paths (A-1-2-5-B and A-3-7-8-B in figure 7.3); in the case of one or two bends, *Z-shaped paths* (all four paths between A and B in figure 7.3).
If only L-shaped paths are allowed, m_ν takes the following form. For each net i between two modules not on a straight line we define $\mu_i = \pm 1$ depending on which L-shaped path is chosen for the net (see figure 7.4). Let $\epsilon_{i\nu} = +1$ along each link ν which net i uses if $\mu_i = 1$, -1 if $\mu_i = -1$ and 0 if link ν is not on the path for net i. The following expression for m_ν is obtained:

$$m_\nu = \sum_i \epsilon_{i\nu}\mu_i + m_\nu(0), \tag{7.2}$$

where $m_\nu(0)$ is the contribution from the straight nets. Vecchi and Kirkpatrick propose a more elaborate expression for m_ν. Define $a_{i\nu}$ as ${\epsilon_{i\nu}}^2$ for each i and ν, then m_ν can also be written as:

$$m_\nu = \sum_i a_{i\nu}\frac{\epsilon_{i\nu}\mu_i + 1}{2} + m_\nu(0). \tag{7.3}$$

Combining eqs. 7.1 and 7.3 yields the following **cost function**:

$$F = \lambda \sum_{i,j} J_{ij}\mu_i\mu_j + \lambda \sum_i h_i\mu_i + \text{constants}, \tag{7.4}$$

with

$$h_i = \sum_{\nu} \epsilon_{i\nu}(2m_\nu(0) + \sum_j a_{j\nu}) \tag{7.5}$$

and

$$J_{ij} = \frac{1}{4} \sum_{\nu} \epsilon_{i\nu}\epsilon_{j\nu}. \tag{7.6}$$

Eq. 7.4 is similar to a Hamiltonian (see section 4.5): h_i can be considered as a ***random magnetic field*** and J_{ij} as the spin-spin interaction in a two-dimensional field of magnetic spins.

An expression for F where Z-shaped paths are allowed is not derived in [VEC83], but it is indicated how such an expression might be obtained.

In the case of L-shaped paths, **transitions** are defined as changes of μ_i in $-\mu_i$ for nets i not on a straight line, in the case of Z-shaped paths they are defined as a random change of the current path for net i to one of the other paths.

Results are presented for a number of randomly generated problems as well as an industrial problem (without any reference to a cooling schedule). The randomly generated problems are obtained by specifying nets at random on a grid of 11×11 modules. The average length of nets is controlled by restricting the positions of the two modules corresponding to each net to an $L \times L$ square ($L < 11$). For 750 randomly generated nets ($L = 3$), the results are displayed in figure 7.5. The improvements obtained by simulated annealing over the random situation are evident.

For the industrial example (consisting of 98 modules) results are reported that compare favourably with those obtained by a sequential *maze-router* [SOU81].

Global routing is also part of the TimberWolf package (Sechen and Sangiovanni-Vincentelli [SEC85], [SEC86a]), where simulated annealing is used to find a global assignment of the nets to the channels in the case of standard cells (see subsection 7.2.2).

Another routing problem, known as channel routing, is considered by Leong *et al.* [LEO85b]. In this problem, a rectangular box is given as well as a multi-terminal net list, each net specifying a set of *terminals* that are to be connected. The terminals are on opposite sides of the

Figure 7.5: Wire crossings corresponding to 750 nets ($L = 3$) placed at random on an 11×11-grid. The thickness of each line is proportional to the number of wires crossing each link. After annealing (**right**) the maximum crossing is reduced from 20 (**left**) to 12 wires/link (reproduced by permission from [VEC83]).

box. The objective is to route the nets in such a way that the number of horizontal tracks in the channel is as small as possible. In figure 7.6 two solutions of a 3-net problem are displayed. Note that nets are not allowed to overlap in the same direction.
Assuming that all multi-terminal nets are broken into two-terminal nets, Leong *et al.* define the *vertical constraint graph* $G_c = (V, E)$ by associating each net with a vertex and defining a directed edge from vertex i to vertex j if the horizontal segment of net i must be placed above that of net j (thus, in the example of figure 7.6, there is a directed edge from 2 to 3). Furthermore, there is said to be a *horizontal constraint* between vertices i and j if the intervals defined by the leftmost and rightmost terminals of nets i and j overlap (in figure 7.6 there are horizontal constraints between 1 and 2 and between 2 and 3).
A partition $\pi = \{V_1, \ldots, V_m\}$ of the vertex set V induces a second graph G_π as follows: each subset V_i of vertices is associated with a vertex $\hat{v}_i$ of G_π and there is a directed edge from $\hat{v}_i$ to $\hat{v}_j$ if there is a directed edge from some $v_k \in V_i$ to some $v_l \in V_j$ in the original graph.

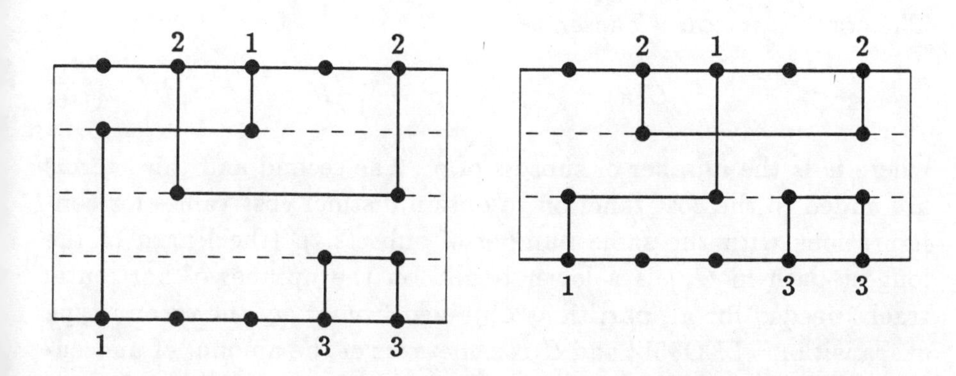

Figure 7.6: Two solutions of a 3-net channel routing problem, in which terminals with corresponding numbers are to be connected. The left-hand solution requires three horizontal tracks, the right-hand only two.

Now, a partition π is said to be *valid* if G_π is acyclic and if there are no horizontal constraints between any two vertices in the same subset (in the problem of figure 7.6, $\{\{1\},\{2\},\{3\}\}$ and $\{\{1,3\},\{2\}\}$ are the corresponding valid partitions). It can be shown [LEO85b] that there is a one-to-one correspondence between valid partitions of the nets and solutions of the routing problem. Moreover, the number of subsets in a partition corresponds to the number of horizontal tracks in the routing solution. Thus, **configurations** can be defined as valid partitions and the routing problem is reduced to finding the valid partition with the smallest number of subsets.
Transitions are defined as:

1. exchanges of two vertices belonging to different subsets in the current partition;

2. changes of a vertex from one subset to another;

3. creation of new subsets by separating a vertex from a subset.

Furthermore, for each transition it should be checked if the resulting partition is again a valid one. In [LEO85b], ways are outlined to carry

out these checks efficiently.
The **cost function** is chosen as:

$$C(\pi) = w^2 + \lambda_p \cdot p^2 + \lambda_u \cdot U, \tag{7.7}$$

where w is the number of subsets of π. The second and third terms are added to the cost function to obtain distinct cost values for configurations with the same number of subsets; p (the length of the longest path in G_π) is a lower bound on the number of horizontal tracks needed for all partitions obtained from π by the second type of transitions [LEO85b] and U is a measure of the amount of unoccupied track, summed over all w tracks. λ_p and λ_c are *weighting factors*, which are experimentally determined.
Using a cooling schedule of the type discussed in section 5.2, a number of well-known problems from literature as well as randomly generated problems are solved. For the problems from literature (including Deutsch' difficult problem [DEU76]), simulated annealing finds the same solutions as those obtained by more traditional algorithms [RIV82] and [YOS82], but in considerably more computation time. On the randomly generated problems, simulated annealing usually outperforms other routing algorithms (which route nets one after the other as in the traditional channel routing algorithms), but the improvement is usually small (for problems with 30-60 two-terminal nets).

In [LEO85a], Leong and Liu consider the *permutation channel routing* problem for multi-terminal nets, a channel routing problem in which some of the terminals are interchangeable. Thus, subsets are given, specifying groups of terminals that are interchangeable, and the problem is to find a permutation of the terminals for which the subsequent channel routing problem can be solved with a minimal amount of horizontal tracks (without, of course, solving the channel routing problem for each permutation). The effect of such a permutation on the subsequent channel routing problem is illustrated in figure 7.7.
A **configuration** of this problem is defined as an assignment of the terminals to the available positions on the border of the channel (*pins*) and **transitions** are defined as pairwise interchanges of permutable terminals. In order to define a **cost function** for an assignment g, the

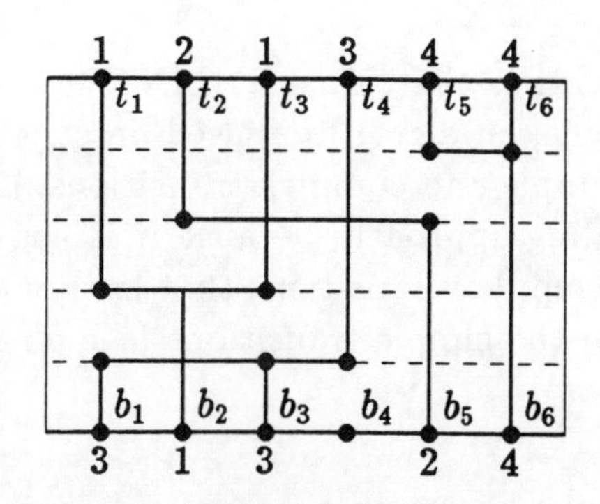

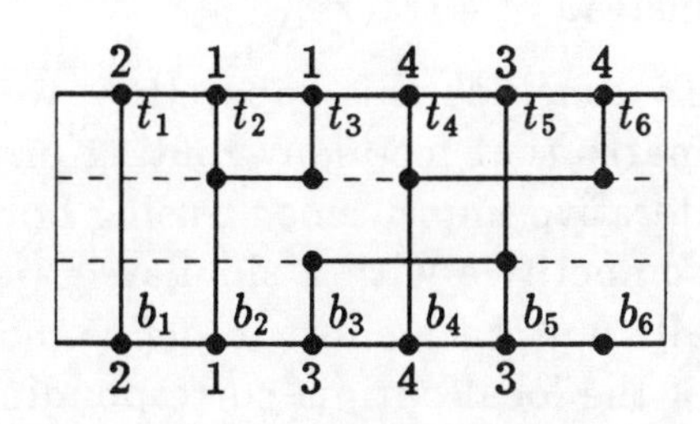

Figure 7.7: Two 4-net channel routing problems; if the terminals on pins $t_1 - t_2$, $t_4 - t_5$, $b_1 - b_5$ and $b_4 - b_6$ are pairwise interchanged, the routing problem can be solved in 2 tracks instead of 4 (reproduced by permission from [LEO85a]).

local density $d_i(g)$ at column i is considered: each net k contributes 1 to the local density of the columns in the interval $[l(k), r(k)]$, where $l(k)$ and $r(k)$ are the left- and right-most pins assigned to net k. The channel density $d(g)$ is defined as the maximum of all local densities. Clearly, the channel density is a lower bound for the number of horizontal tracks in the optimal solution of the subsequent channel routing problem; in figure 7.7, the left-handed assignment has channel density 3, the right-handed has density 2.
The cost of an assignment is now given by

$$C(g) = \lambda d^2(g) + \frac{1}{l} \sum_{i=1}^{l} d_i{}^2(g), \tag{7.8}$$

where l is the total number of columns and λ a weighting factor.
Using again a simple cooling schedule (see section 5.2), a number of randomly generated problems as well as some channel routing examples from literature are solved. Simulated annealing usually finds slightly smaller channel densities than iterative improvement; compared with a result previously obtained by using a more traditional algorithm [KOB84], simulated annealing finds a 50% smaller channel

density (the subsequent channel routing problem can be solved in 3 instead of 4 tracks).

To conclude this subsection, we mention that Linsker [LINS84] remarks that for many routing problems, an algorithm based on strict iterative improvement, using both simple and complex transitions, is competitive with a simulated annealing approach, because it is usually not much effort to define more complex transitions that lead out of the local optima corresponding to the simple transitions (see also chapter 1).

7.2.4 Other applications in computer-aided circuit design

In this subsection applications of simulated annealing to *logic minimization* (Fleisher *et al.* [FLE84], [FLE85], Lam and Delosme [LAM86b], Gonsalves [GON86]), *digital filter design* (Catthoor *et al.* [CAT85], [CAT86]), *delay reduction* (Pincus and Despain [PIN86]), *floorplanning* (Otten and Van Ginneken [OTT84], Wong and Liu [WON86b]), *test pattern generation* (Ligthart *et al.* [LIG86]), *behavioural testing* (Distante and Piuri [DIS86]), *folding* (Moore and de Geus [MOO85], Wong *et al.* [WON86a]) and *two-dimensional compaction* (Zeestraten [ZEE85], Mosteller [MOS86], Osman [OSM87]) are discussed.

When designing electronic circuits, one is often confronted with optimization problems concerning the implementation of *Boolean functions*, i.e. functions which map binary variables into binary variables [BRE72]. Such a function can always be written as the OR-sum of AND-products of its variables. Take for example a function f of three binary variables x_0, x_1 and x_2, which outputs 1 if at least two of the variables are 1 (in circuit theory, f is known as the *carry output of a full adder*), then f can be represented as the Boolean sum of four products:

$$f(x_0, x_1, x_2) = x_0 \cdot x_1 + x_0 \cdot x_2 + x_1 \cdot x_2 + x_0 \cdot x_1 \cdot x_2. \tag{7.9}$$

Often it is possible to represent a Boolean function as a sum of products in several ways and one is then interested in finding the represen-

tation with the smallest number of products (in eq. 7.9, for example, the last product in the sum can be deleted without changing the functionality of f). This problem is usually referred to as *logic minimization.* Many approximation algorithms to solve this problem have been published in recent years [BRAY84]. In [FLE85], Fleisher *et al.* propose simulated annealing as an algorithm to solve logic minimization and related problems. One of these related problems is considered in more detail (see also [FLE84]): given a Boolean function f, the problem is to decompose f into two functions u and v, such that f is the *exclusive-OR* sum of u and v (the exclusive-OR of two binary variables x_0 and x_1 is given by $x_0 \cdot \overline{x}_1 + \overline{x}_0 \cdot x_1$). We are interested in finding the decomposition for which the numbers of products in the sum-of-products representations of u and v are minimal.
Each **configuration** is given by a Boolean function u, from which v is uniquely determined via $v = f \oplus u$ ($\oplus$ denotes exclusive-OR sum, $v = f \oplus u$ follows from $f = u \oplus v$), **transitions** are defined as small changes in the function u (e.g. changing a variable in the representation of u to its negation). To have an exact cost value for each configuration would necessitate the solution of two logic minimization problems (for u and v) for each configuration, which would result in prohibitive computation times. Thus, for each configuration, a simple and fast algorithm is used to find approximately minimal sum-of-products representations for u and v. If p_u and p_v are the amounts of products in the representations found for u and v, respectively, then the **cost function** is defined as [FLE84]

$$C(u,v) = p_u + p_v + \lambda(| p_u - p_v |), \tag{7.10}$$

where λ is a weighting factor. The difference term in eq. 7.10 is added to bias solutions in favour of balanced decompositions.
Fleisher *et al.* [FLE85] report results for 14 Boolean functions of 2 to 18 variables; in the largest example (18 variables), simulated annealing reduces the 11 products in the representation of f to a total of 6 products in the representations of u and v.
An application in *multi-level logic minimization* is described by Lam and Delosme [LAM86b]. Here, one is interested in finding the smallest representation of a Boolean function, not necessarily as a sum of products. The function f given by eq. 7.9, for

example, can also be written as:

$$f(x_0, x_1, x_2) = x_0 \cdot x_1 + (x_0 + x_1) \cdot x_2. \tag{7.11}$$

Configurations correspond to Boolean expressions and are represented by means of graphs, **transitions** are implemented by logic operations on these graphs. Results of the same quality as those of a human designer are reported.
In a similar way, simulated annealing is used by Gonsalves [GON86] to find a minimum representation of a Boolean function in terms of *NORs* and *NANDs* (inverted OR-sums and AND-products, respectively), and for transforming a given representation of a Boolean function in a representation consisting only of elements of a given set of Boolean operators. A five percent improvement over the existing method is reported.

An application of simulated annealing in ***digital filter design*** is reported by Catthoor *et al.* [CAT85]. If all inputs of a digital filter are 0, then the outputs should eventually converge to 0 as well. Due to the finite precision of arithmetic operations in a filter, convergence to 0 is not observed in practice, instead one observes ***zero-input limit cycles.*** By selecting an appropriate ***quantization mechanism***, the maximum amplitude of these cycles can be controlled. One is now interested in finding the quantization for which the maximum amplitude is as small as possible.
The corresponding optimization problem can be formulated as an integer linear programming problem, where a function of integer variables is to be minimized, given a number of linear constraints on the variables. Simulated annealing is proposed as an optimization technique to solve this problem. Part of the employed cooling schedule was already discussed in section 5.3 (see Table 5.1). Simulated annealing is compared with an approach based on strict iterative improvement and from the experimental results it is concluded that, in order to obtain the same quality of results with iterative improvement as with simulated annealing, the former technique needs 20 to 60% more computation time. Thus, for this problem simulated annealing compares favourably with iterative improvement.
Some other applications in digital filter design, all related to the investigation of finite word-length effects, are reported by Catthoor *et*

al. in [CAT86].

Besides area, *timing* is an important aspect in the design of an electronic circuit: the circuit is required to produce its output within a specified time. Meeting such a time or *delay* constraint can usually be realized by increasing the sizes of the transistors in the circuit. Thus, there is a trade-off between delay and area and one is asked to minimize delay with a constraint on total transistor area. In [PIN86], Pincus and Despain use simulated annealing to solve this optimization problem. A **configuration** is simply an assignment of sizes to the transistors, **transitions** are implemented by randomly perturbing a transistor-size and the cost function includes a penalty for exceeding the specified delay and a term relating to the total size of the transistors. A mechanism, similar to the one described by Greene and Supowit [GREE84], [GREE86] is used to improve the efficiency of the generation mechanism for transitions.

The *floorplan design* problem is considered by Otten and Van Ginneken [OTT84] and Wong and Liu [WON86b]. It is a generalization of the placement problem (see subsection 7.2.2): the problem is again to place rectangular modules such that total area and length of interconnections are minimum. However, modules do not have a fixed size, but may assume any shape permitted by their *shape constraints.* Such a constraint bounds the size of a module in one dimension as a function of the size in the other dimension. A special case of this problem was already discussed in subsection 7.2.2: the placement of modules with varying width and height but fixed area.
Wong and Liu [WON86b] restrict themselves to floorplans corresponding to *slicing structures.* A slicing structure is a rectangle dissection that can be obtained by recursively cutting rectangles into smaller rectangles, see figure 7.8. To each slicing structure there corresponds uniquely a *normalized Polish expression* [WON86b]. A Polish expression of length $2n-1$ is a sequence $\alpha_1, \ldots, \alpha_{2n-1}$ of elements from $\{1, 2, \ldots, n, +, \star\}$, such that every i in $\{1, 2, \ldots, n\}$ appears exactly once in the sequence and such that for all $k \in \{1, 2, \ldots, 2n-1\}$ the sequence $\alpha_1, \ldots, \alpha_k$ contains more operands (the numbers $1, \ldots, n$) than operators $(+, \star)$. A Polish expression is said to be normalized iff there are no consecutive $\star$'s or $+$'s in the sequence. If we associate the

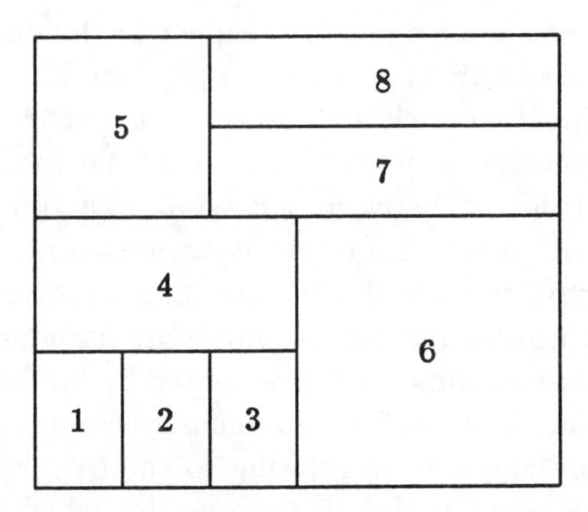

Figure 7.8: An example of a slicing structure.

operator + to 'on top of' and $\star$ to 'next to', we can think of the Polish expression as the postfix notation for the 'arithmetic expression' associated to a floorplan. Thus, $12\star3\star4+6\star578+\star+$ is the normalized Polish expression corresponding to the floorplan in fig 7.8. If the problem is to place n modules, Wong and Liu define **configurations** as normalized Polish expressions of length $2n-1$. **Transitions** are defined as

1. exchange of two consecutive operands in the current sequence;
2. interchange of the operators $\star$ and + in a subsequence of operators (e.g. a change of $\star+\star+\star$ in $+\star+\star+$);
3. exchange of an operand with an adjacent operator in the current sequence.

The third type of transition may violate the requirement that the resulting sequence is again a normalized Polish expression, but there is an $\mathcal{O}(1)$-time algorithm to check this [WON86b]. Finally, the cost function is defined as

$$C = A + \lambda W, \tag{7.12}$$

where λ is a weighting factor. The total area A is obtained in the following way. Given the shape constraints of all n modules the floorplan

is composed of, it is straightforward to compute the shape constraint of the floorplan itself. The total area A is then found by choosing the point on the shape constraint of the floorplan that minimizes total area. Once the shape of the floorplan is fixed, the shapes of all modules can also be determined. This gives the total estimated length of connections W: for each connection, connecting modules i and j, its length is estimated by the distance between the centers of modules i and j in the current floorplan.
Using a 'simple' cooling schedule (see section 5.2), Wong and Liu apply simulated annealing to a number of randomly generated problems, varying in size from 15 to 40 modules. CPU-times on a PYRAMID-computer range from 1 to 13 minutes. For all problems the area of of the final floorplan is less than 4% larger than the total area of all modules (a theoretical lower bound for the global minimum).
The same problem is treated by Otten and Van Ginneken in [OTT84], but configurations, transitions and cost function are chosen differently. **Configurations** are represented by means of *point configurations* - points represent modules and are obtained in such a way that the relative positions of the modules in a floorplan are preserved. Rather than taking the points themselves as configurations, the orderings along both axes implied by a point configuration are defined as configurations. **Transitions** are implemented by pairwise exchanges of points along either of the two axes. The **cost function** is based on the total estimated wire length (calculated by using geometrical information obtained from the distance between the points). The employed cooling schedule was discussed in section 5.3; unfortunately no results are reported.

Simulated annealing is used by Ligthart *et al.* [LIG86] as optimization technique for a problem which arises when generating a set of *test patterns* for a special kind of circuit, called *Programmable Logic Arrays* (PLA's). Without referring to the theory or concepts of test pattern generation, the optimization problem can be described as follows. A $k \times n$-matrix $A = (a_{ij})$ is given, whose elements are either '0', '1' or '-', the latter denoting a *don't care*, meaning that the corresponding element can be either 0 or 1. To each row of A, a set S_i of binary sequences, called the select set, is associated, which consists of all 2^{s_i} sequences corresponding to row i (s_i is the number of don't-cares in row i). For example, the select set of $(1,-,0,-)$ is given by $\{(1,0,0,0),(1,0,0,1),(1,1,0,0),(1,1,0,1)\}$. The problem is to find a set of sequences $\{u_i\}$, $u_i \in S_i$, **one** from each select set, in such a way that for all u_i the *Hamming distance* to the elements of all other select sets S_j $(j \neq i)$ is at least 2, i.e.

$$\forall i,j \in \{1,\ldots,k\},\ i \neq j : \min_{s_l \in S_j} d_H(u_i, s_l) \geq 2, \tag{7.13}$$

where the Hamming distance $d_H(u,v)$ of two binary sequence u and v is defined as

$$d_H(u,v) = \sum_i u_i \oplus v_i. \tag{7.14}$$

Of course, such a set of sequences u_i, $i = 1,\ldots,k$ may not exist. In that case, an extra column is added to the matrix A and the elements of the column are chosen 0 or 1, in such a way that the search for a set of sequences, satisfying eq. 7.13, is facilitated. The optimization problem therefore consists of finding such a set of sequences with the smallest amount of extra columns for the matrix A.
Each **configuration** is characterized by a choice of sequences u_i, one from each S_i, and **transitions** are realized by replacing at random an u_i by another element of S_i. The **cost function** is defined as

$$C = \frac{\gamma \cdot \max(|M_<|,|M_>|) + |M_>| \cdot \min(|M_<|,|M_>|)}{|M_<| + |M_>|}, \tag{7.15}$$

where

$$M_< = \{d_{ij} \mid d_{ij} < 2\},\ M_> = \{d_{ij} \mid d_{ij} > 2\} \tag{7.16}$$

and the distances d_{ij} are defined as

$$d_{ij} = \begin{cases} \min_{s_l \in S_j} d_H(u_i, s_l) & \text{if } i \neq j \\ 2 & \text{if } i = j. \end{cases} \tag{7.17}$$

The positive weighting factor γ is defined in such a way that $C = 0$ if $|M_<| = 0$ (which corresponds to a solution of the problem). Thus, the problem can be solved by repeatedly optimizing the cost function for increasing values of k, until finally a solution is found for which $C = 0$.

With a cooling schedule of the type discussed in section 5.2, results are obtained for a number of problems from the literature which are slightly better than previously published results but in considerably larger amounts of computation time.
Another application in testing of VLSI circuits is mentioned in [DIS86], where Distante and Piuri report on the construction of an optimal *behavioural test* procedure using simulated annealing.

Moore and de Geus [MOO85] as well as Wong *et al.* [WON86a] consider *folding* of PLA's. Our description of the problem closely follows the terminology of Wong *et al.* An $m \times n$-matrix $A = (a_{ij})$ is given, consisting of 0's and 1's and the problem is to find a permutation π of the rows that implies a folding of minimal size. A folding with respect to π is defined as a partition of the columns of A into sets of mutually *foldable* columns, where two columns c_i and c_j are foldable iff

$$u_i^\pi < l_j^\pi \text{ or } u_j^\pi < l_i^\pi, \tag{7.18}$$

where

$$l_i^\pi = \min\{\pi(k) \mid a_{\pi(k)i} = 1,\ k = 1,\ldots,m\} \tag{7.19}$$

and

$$u_i^\pi = \max\{\pi(k) \mid a_{\pi(k)i} = 1,\ k = 1,\ldots,m\}. \tag{7.20}$$

The size of a folding is defined as the number of sets in the partition. The aforementioned problem is referred to as *multiple-column folding*; *simple* or *two-column folding* implies that the sets of a partition can contain at most two columns. In [WON86a], simple folding is treated as a special case of multiple column folding.
Clearly, for a given permutation π, there are many different foldings (i.e. partitions of the set of columns). Fortunately, the size of the smallest folding with respect to π can be easily determined without constructing the folding itself in the following way. Associate to each column c_j an interval $I_j^\pi = [l_j^\pi, u_j^\pi]$, where l_j^π and u_j^π are given by eqs. 7.19 and 7.20, respectively. Let the density d_i^π of a row r_i, $i \in \{1,\ldots,m\}$, be the number of intervals it intersects and let the density d^π be the maximum density of all rows. It can be shown that this density is equal to the size of the smallest folding with respect to π [WON86a]. Thus, **configurations** can be represented by permutations of the rows and the folding problem is reduced to finding

a permutation π, such that the corresponding density d^π is minimal. **Transitions** are defined by exchanging two rows in the current permutation, the **cost function** is chosen as

$$C(\pi) = (d^\pi)^2 + \frac{\lambda}{m} \sum_{i=1}^{m} (d_i^\pi)^2, \tag{7.21}$$

where λ is a weighting factor.
Wong *et al.* use a cooling schedule of the type discussed in section 5.2 and compare simulated annealing to an optimization algorithm [SKR85] and a well-known approximation algorithm [DEM83]. In the case of multiple folding, simulated annealing solves all problems under consideration to optimality and consistently outperforms the approximation algorithm. CPU-times on a PYRAMID-computer vary from 0.7 seconds for a 6×10-matrix to 222 seconds for a 70×50-matrix.
In [MOO85], Moore and de Geus treat the same problem, but configurations, transitions and the cost function are chosen differently. For each problem, the cooling schedule is determined by consulting a knowledge base, acquired by performing a large number of experiments with extensive diagnostics. For each of these experiments, the best cooling schedule is determined empirically.
Results are reported which are on average 10% better in quality of solution than those obtained by the aforementioned approximation algorithm [DEM83] in two to three times as much computation time.

Zeestraten [ZEE85], Mosteller [MOS86] and Osman [OSM87] consider the *two-dimensional compaction* problem. Compaction can be viewed as a special kind of placement: the topology of modules and connections is given and the problem is to find absolute positions of modules and connections for which the total area is as small as possible and minimum distance constraints between modules and connections are satisfied. Examples of an initial configuration and the one after compaction are displayed in Figs. 7.9a and 7.9d, respectively. Usually, the problem is solved by first compacting modules and connections in one dimension, followed by compaction in the other dimension. Figs. 7.9b and 7.9c clearly illustrate that this approach leads to non-optimal results: compared with the optimum configuration in figure 7.9d, the configurations in Figs. 7.9b and 7.9c take a considerably larger area.

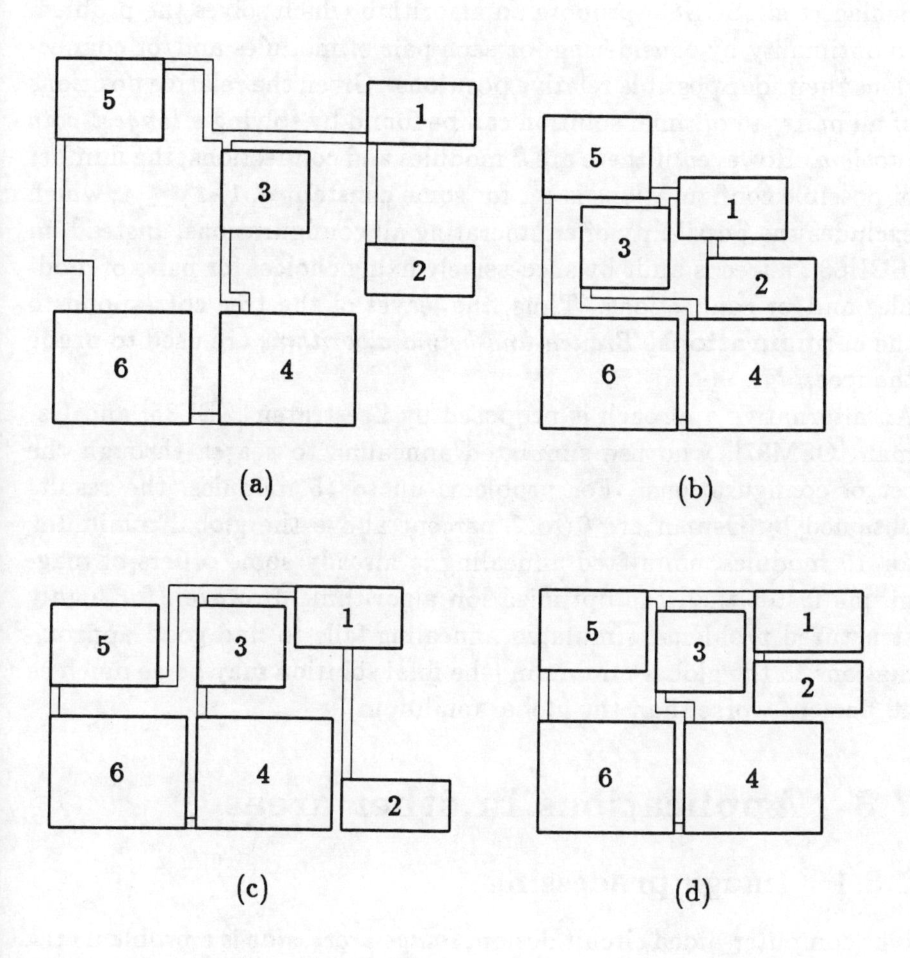

Figure 7.9: A compaction problem; (a) initial configuration, (b) configuration after x- followed by y-compaction, (c) configuration after y- followed by x-compaction, (d) configuration after 2-dimensional compaction (reproduced from [ZEE85]).

Note that the positions of modules relative to each other can change: whereas module 2 is above module 4 in figure 7.9b, it is to the left of module 4 in figure 7.9c.
Schlag *et al.* [SCH83] propose an algorithm which solves the problem to optimality by considering for each pair of modules and/or connections their four possible relative positions. Given the relative positions of all pairs, an optimal solution can be found by solving a ***longest-path problem***. However, if there are k modules and connections, the number of possible configurations is p^k, for some constant p, $1 < p \leq 4$, which excludes the possibility of enumerating all configurations. Instead, in [SCH83], a tree is built by successively fixing choices for pairs of modules and/or connections. Thus, the leaves of the tree correspond to the **configurations**. ***Branch-and-bound algorithms*** are used to prune the tree.
An alternative approach is proposed by Zeestraten [ZEE85] and Osman [OSM87], who use simulated annealing to search through the set of configurations. For problems up to 15 modules, the results obtained by Osman are 0 to 7 percent above the global minimum; for 15 modules, simulated annealing is already some orders of magnitude faster than the optimization algorithm. However, for highly structured problems, simulated annealing fails to find good approximations to the global minimum (the final solution may be as much as 40 percent worse than the global minimum).

7.3 Applications in other areas

7.3.1 Image processing

Like computer-aided circuit design, ***image processing*** is a problem area in which many large-scale optimization problems occur, for which either no or otherwise unsatisfactory algorithms exist. The first paper dealing with simulated annealing in this context was published by Geman and Geman [GEM84], in which a generalization of simulated annealing is used to find a maximum ***posterior distribution*** for a degraded image. Their approach is briefly discussed hereinafter. A simplified version of this approach can be found in [WOL85]. Gidas' approach to image processing problems [GID86] can be viewed as an elegant extension based on ***renormalization group*** ideas.

An ***image*** X is considered as a realization of $n \times n$ random variables $\mathbf{X}(i,j)$ on an $n \times n$-grid, corresponding to the ***pixels*** of the image. Geman and Geman show that if the stochastic variable $\mathbf{X}$ is a *Markov random field*, i.e. if $\mathbf{X}$ satisfies

$$Pr\{\mathbf{X}(i,j) = X(i,j) \mid \mathbf{X}(k,l) = X(k,l),\ (k,l) \neq (i,j)\} =$$

$$Pr\{\mathbf{X}(i,j) = X(i,j) \mid \mathbf{X}(k,l) = X(k,l),\ (k,l) \in \mathcal{R}_{ij}\}, \qquad (7.22)$$

where $\mathcal{R}_{ij}$ denotes the neighbourhood of pixel (i,j), then the probability distribution of $\mathbf{X}$ is given by a Gibbs (Boltzmann) distribution,

$$Pr\{\mathbf{X} = X\} = \frac{1}{Z} \exp\left(\frac{-C(X)}{c}\right), \qquad (7.23)$$

where Z is a normalizing constant and $C(X)$ the cost of an image, suitably defined (the cost consists of terms for each pixel, each term depending only on the values of $\mathbf{X}$ in the neighbourhood of the pixel). Note that eq. 7.22 is a generalization of

$$Pr\{\mathbf{X}(k) = i \mid \mathbf{X}(j) = X_j,\ j < k\} =$$

$$Pr\{\mathbf{X}(k) = i \mid \mathbf{X}(k-1) = X_{k-1}\}, \qquad (7.24)$$

which characterizes a Markov chain.

Suppose that instead of X a corrupted image Y is observed, i.e. $Y = F(X)$, a distortion of the noise-free image X, and the problem is to find X, given Y (for this reason, this type of problem is known as an ***inverse problem***[2]). A ***Bayesian approach*** to this problem defines the image X for which $Pr\{\mathbf{X} = X \mid \mathbf{Y} = Y\}$ is maximal as the solution to this problem. It is well known that

$$Pr\{\mathbf{X} = X \mid \mathbf{Y} = Y\} = \frac{Pr\{\mathbf{Y} = Y \mid \mathbf{X} = X\} \cdot Pr\{\mathbf{X} = X\}}{Pr\{\mathbf{Y} = Y\}}, \qquad (7.25)$$

[2] In a recent paper [ČER86b] Černy argues that such problems can be solved by iterating between the inverse problem and the conjugated ***direct*** problem (given X, find Y), where the inverse problem is formulated as an optimization problem with cost function given by the analogue of ***enthalpy***, instead of energy.

where $Pr\{\mathbf{X}\}$ and $Pr\{\mathbf{X} \mid \mathbf{Y}\}$ are called the ***prior*** and ***posterior distributions***, respectively. Since the denominator in eq. 7.25 does not depend on X, we are interested only in maximizing the numerator of eq. 7.25. Furthermore, Geman and Geman show that under certain assumptions, $Pr\{\mathbf{X} \mid \mathbf{Y}\}$ is again, as $Pr\{\mathbf{X}\}$, a Boltzmann distribution. Thus, a Markov chain (or, more precisely, a Markov Random Field) can be set up to approach this distribution and by lowering the value of the control parameter we approach the image with the lowest cost, i.e. the "most probable" image.
For each value of c, the procedure proposed by Geman and Geman runs as follows (see also [WOL85]). Let X be the current estimate of the undisturbed image and suppose each $\mathbf{X}(i,j)$ can take κ values. For binary images, for example, κ would be 2. For a randomly chosen pixel of the image, the current value is replaced by one of the κ possible alternatives to obtain κ new images $X_1, \ldots, X_\kappa$. Next, the probabilities $Pr\{\mathbf{Y} = Y \mid \mathbf{X} = X_i\}$, depending on the distortion mechanism F, and $Pr\{\mathbf{X} = X_i\}$, given by eq. 7.23, are calculated. The new pixel value is now obtained from the κ alternatives according to the following probability distribution

$$p_i = \frac{Pr\{\mathbf{Y} = Y \mid \mathbf{X} = X_i\} \cdot Pr\{\mathbf{X} = X_i\}}{\sum_{j=1}^{k} Pr\{\mathbf{Y} = Y \mid \mathbf{X} = X_j\} \cdot Pr\{\mathbf{X} = X_j\}}, \quad i = 1, \ldots, \kappa. \quad (7.26)$$

Geman and Geman show that

1. for each value of the control parameter, the sequence of images generated according to eq. 7.26 asymptotically converges to the probability distribution given by eq. 7.23 (cf. convergence to the stationary distribution of a homogeneous Markov chain, as discussed in section 3.1), provided each pixel continues to be visited often enough;

2. for a decreasing sequence of values of the control parameter $\{c_k\}$, satisfying

$$c_k \geq \frac{\Gamma}{\log k}, \quad (7.27)$$

for large enough k and some constant Γ, the sequence of images asymptotically converges to a uniform distribution on the set of images with minimum cost.

Rothman [ROT85], [ROT86a] (who considers inverse problems in *seismology*), Geman and Geman [GEM84], Ripley [RIP86] and Wolberg and Pavlidis [WOL85] report results for several kinds of distortions, which are of good quality and sometimes superior to results obtained by previously available algorithms. Research in this area, however, is still in its early stages and not much more can be said than that the approach seems to be promising.
Another interesting application of the simulated annealing algorithm to image processing is presented by Güler *et al.* They apply the algorithm for delineating geological faults on radar images (in particular, images of the San Andreas fault in California). Their approach consists of two steps. The first step is the row-by-row detection of line elements or points at which the grey level statistics on a radar image, taken by a satellite, changes abruptly. In the second step, these line elements and points are connected into lines using a priori geological information and using a *line restoration* algorithm based on simulated annealing.
Other applications of simulated annealing in image processing are reported by Carnevalli *et al.* [CARN85], Sontag and Sussmann [SON85] (*image segmentation*) and Smith *et al.* [SMI85], [SMI83] (*reconstruction of coded images*). For a discussion of these problems in a wider context, the reader is referred to [RIP86].

7.3.2 Boltzmann machines

Revolutionary computer architectures such as *connection machines* [HIL85] are based on connectionist models [FAH83], [FELD85]. These models incorporate massive parallelism as well as the hypothesis that adaptive adjustment of the strength of inter-neural connections is what produces learning in the neural network of the human brain. The *Boltzmann machine* (see also section 8.1.3), introduced in 1983 by Hinton *et al.* [HIN83], [HIN84], is a self-organizing architecture that can be viewed as a novel approach to connectionist machines. The architecture is based on a *neural network* [HOP82] and incorporates a distributed knowledge representation and stochastic computing elements, the latter being the reason why it is discussed here. Salient features of the Boltzmann machine are its memory, associative and inductive capabilities [AAR86c], [ACK85].

The Boltzmann machine is thought to consist of a massively parallel network of simple *logical units* that can be in two states, either "on" or "off". The set of logical units consists of three subsets, those of *input units* (V_i), *output units* (V_o) and *hidden units* (V_h). A Boltzmann machine has the ability to learn a Boolean relation defined on the input and output units. Learning takes place by examples. This is done by externally forcing (or clamping) the states of the input and output units to specific binary input-output combinations for which the Boolean relation holds.

The units are mutually connected by bidirectional links. A *connection strength* s_{ij} is associated with each link, representing a quantitative measure for the hypothesis that the two units interconnected by the link are both "on" ($s_{ij} = 0$ if units i and j are not connected). A self-connection strength s_{ii} is used to introduce a threshold for each unit. If $s_{ij} \gg 0$ then it is considered very desirable that both units i and j are "on", if $s_{ij} \ll 0$ it is considered very undesirable.

Let n be the total number of units. Then, a configuration k of the system can be uniquely represented by an n-dimensional binary configuration vector $\mathbf{r}_k \in \{0,1\}^n$. The i-th component $r_k(i)$ of the configuration vector $\mathbf{r}_k$ denotes the state of the i-th unit v_i in configuration k, 1 corresponds to "on" and 0 with "off".

The *consensus* C_k assigns a real number to a configuration k, which is a quantitative measure of the amount of consensus in the system with respect to the set of underlying hypotheses and is defined as

$$C_k = \sum_{i=1}^{n} \sum_{j=i}^{n} s_{ij}\, r_k(i)\, r_k(j). \tag{7.28}$$

If a configuration k is transformed into a configuration k' by changing the state of an arbitrary unit i ($0 \rightarrow 1$ or vice versa) then the corresponding difference in the consensus $\Delta C_{kk'}$ is given by

$$\Delta C_{kk'} = (r_{k'}(i) - r_k(i))\Big(\sum_{j=1, j\neq i}^{n} s_{ij} r_k(j) + s_{ii}\Big). \tag{7.29}$$

From eq. 7.29 it is apparent that the effect on the consensus of the system by changing the state of vertex v_i is completely determined by the states of the vertices v_j that are connected to v_i and by the corresponding connection strengths. Consequently, the differences in consensus $\Delta C_{kk'}$ can be computed locally, thus allowing parallel execution.

The *learning algorithm* associated with the Boltzmann machine [AAR86c], [ACK85] starts off by setting all connection strengths equal to zero. Next, a sequence of learning cycles is completed, each cycle consisting of two phases. In the first phase a set of learning examples is clamped on the input and output units (clamped situation) and for each learning example the consensus is optimized based on the current set of connection strengths by adjusting the configuration of the system. In the second phase all units in the Boltzmann machine are left free (free-running situation) and again consensus optimization is carried out. In between subsequent learning cycles the connection strengths are adjusted using statistical information obtained from the two phases of the previous learning cycle. This process is continued until the average change (over a number of learning cycles) of the connection strengths is zero. After completion of the learning algorithm the Boltzmann machine is able to complete a partial example (i.e. a situation where only a subset of the input and output units is clamped) by maximizing the consensus.
By consensus optimization, within the scope of the learning algorithm, we mean modification of the free-running units such that configurations with a larger consensus occur with higher probabilities. Thus, the goal is to generate a distribution of configurations rather than one particular configuration with probability one.

The learning algorithm itself is based on the minimization of an *asymmetric divergence* G, which is a quantitative measure of the distance between two probability distributions P and P', that are related to the probability distributions of the states of the input and output units in the free-running and clamped situations, respectively [AAR86c], [ACK85]. The objective of the learning algorithm now can be rephrased as: **minimize G by changing the connection strengths**.

If the transitions of free-running units (both in the clamped and the free-running situation) are modelled stochastically such that a stable pattern (equilibrium) is generated in which a configuration k with a consensus C_k occurs with a probability $q_k(c)$ given by

$$q_k(c) = q_0(c) \exp\left(\frac{C_k - C_{max}}{c}\right), \tag{7.30}$$

where C_{max} denotes the maximal value of the consensus of the system for a given set of connection strengths, c the control parameter and $q_0(c)$ a normalization constant such that $\sum_{k=1}^{|\mathcal{R}|} q_k(c) = 1$, then we have that

1. configurations corresponding to a higher consensus have a larger probability to occur than configurations with a lower value of the consensus, and

2. the partial derivative of the asymmetric divergence G with respect to s_{ij} can be written as [ACK85]

$$\frac{\partial G}{\partial s_{ij}} = \frac{<p'_{ij}> - <p_{ij}>}{c}, \tag{7.31}$$

where $<p_{ij}>$ denotes the expected value of the probability that both units i and j are "on" in the clamped situation and $<p'_{ij}>$ the expected value of the probability that both units are "on" in the free-running situation, when equilibrium is reached.

From 1. it follows that the Boltzmann machine favours configurations with larger consensi. From 2. it follows that to minimize G it suffices to collect statistics on $<p_{ij}>$ and $<p'_{ij}>$ and to change the connection strength in proportion to the difference between the expected values (gradient method), i.e.

$$\Delta s_{ij} = \eta\ (<p_{ij}> - <p'_{ij}>), \tag{7.32}$$

where η denotes a constant.
The learning algorithm can now be described more precisely as follows: the algorithm starts off with all connection strengths set to zero

(tabula rasa). Next, a number of learning cycles is completed by repeatedly changing the system from the clamped to the free-running situation and collecting statistics on $< p_{ij} >$ and $< p'_{ij} >$. In between subsequent learning cycles the connection strengths are adjusted until G is minimal.

To realize the probability distribution, given by eq. 7.30 the simulated annealing algorithm is used. The generation probability is chosen uniformly over all units that can change their state. The acceptance probability is chosen as

$$A_{kk'}(c) = (1 + e^{-\Delta C_{kk'}/c})^{-1}, \tag{7.33}$$

where $\Delta C_{kk'}$ is given by eq. 7.29. Clearly, within the learning algorithm for the Boltzmann machine the simulated annealing algorithm is only used to generate a stable pattern reflecting the properties of eq. 7.30, which means that in this case the limit $c \downarrow 0$ is not approximated, but that the algorithm is terminated at some value of c for which the acceptance ratio $\chi(c)$ still differs significantly from 0.
To test the results obtained by the learning algorithm during the learning phase a *test algorithm* is introduced by Aarts and Korst [AAR86c]. In the test phase only the input units are clamped. For a given input the consensus is maximized using the connection strengths obtained from the learning phase. Here both the hidden and the output units are considered as free units. After optimization of the consensus the test can be completed by comparing the resulting output, given by the states obtained for the output units, with the "expected" output. To obtain a unique output again the simulated annealing algorithm is used but now consensus optimization is carried out until a configuration with (near-)maximum consensus is obtained occurring with a probability close to one (approximation of the limit $c \downarrow 0$). Thus, in the test algorithm, consensus optimization is carried out along the lines of the simulated annealing algorithm as described in the previous chapters.

7.3.3 Miscellaneous applications

Finally, applications of simulated annealing in *numerical analysis* (Armstrong [ARM84]), *biology* (Lundy [LUN85]), *magnetics* (Lyberatos *et al.* [LYB85]), *materials science* (Wooten *et al.* [WOOT85]), *game theory* (Van Laarhoven *et al.* [LAA88a], Wille [WIL87b]), *code design* (Beenker *et al.* [BEE85], Bernasconi [BER87], El Gamal *et al.* [ELG87]), and *wheel balancing* (Masarik [MAS85]) as well as some other applications mentioned by Uhry [UHR86] are briefly described:

1. given a square matrix A, its *bandwidth* is defined as the largest row bandwidth: the number of columns in a row from the first non-zero up to and including the diagonal. In [ARM84], Armstrong presents an algorithm which finds a permutation of the rows, minimizing the bandwidth of a matrix, and uses simulated annealing as one of the optimization techniques. Satisfactory results are reported, in terms of both bandwidth and computation time.

2. the *evolutionary tree* problem is considered by Lundy [LUN85]: a set of n populations is given, which was obtained as follows. Originally there was one population, say A, which split into two smaller ones, say B and C. B and C did not interbreed and each split again into two different populations. The process continued and finally resulted in the given set of n populations. The problem now is to find the "best" tree representation, where certain criteria from biology serve as guidelines to determine what "best" is. The optimization problem can be shown to be equivalent to finding the minimum length *Steiner tree* for n elements. Simulated annealing is reported to compare favourably with an iterative improvement approach in terms of quality of solution.

3. in [LYB85], Lyberatos *et al.* show that simulated annealing can be used to calculate optimal configurations of a system of interacting fine ferromagnetic particles; the *ac magnetic field* is shown to play a role similar to that of the control parameter in simulated annealing.

4. in [WOOT85], Wooten *et al.* use simulated annealing to construct realistic random-network models of amorphous Si with periodic boundary constraints.

5. the *football pool* problem is considered by Wille [WIL87b]: given the set V_3^n of all n-tuples $\mathbf{x} = (x_1, \ldots, x_n)$, with $x_i \in \{0,1,2\}$, find σ_n, the minimal size of a subset W of V_3^n, such that for each element $\mathbf{x}$ in V_3^n there is at least one element $\mathbf{y}$ in W, satisfying $d_H(\mathbf{x},\mathbf{y}) \leq 1$ (d_H denotes Hamming distance, see eq. 7.14). σ_n can be viewed as the minimal number of forecasts in a football pool of n matches, where at least one forecast has at least $n-1$ correct results. By using simulated annealing, is is shown that $\sigma_6 \leq 74$, whereas the previous best upper bound was 79. In [LAA88a], this bound is improved to 73, and new upper bounds are reported for σ_7 and σ_8.

6. in [ELG87], El Gamal *et al.* consider the problem of representing the set of all 2^L *binary sequences* of length L by a much smaller subset of 2^K code words ($K \ll L$) in such a way that the average Hamming distance between each of the 2^L sequences and its nearest code word is minimal. Results obtained by using simulated annealing are reported to be very encouraging. Other topics in coding theory, the search for *binary sequences with maximally flat amplitude spectrum* and for *sphere-packing bounds for binary sequences* are discussed by Beenker *et al.* [BEE85] and Bernasconi [BER87], and by El Gamal *et al.*, respectively.

7. Uhry [UHR86] discusses applications of simulated annealing to *optimization of cutting patterns* of irregularly shaped pieces, e.g. in the garment industry, to *vehicle routing* under capacity and time-window constraints (the monthly routing of dustbin lorries in Grenoble (France) is nowadays found by using simulated annealing) and to the problem of *finding good 3-dimensional meshes* to be used in finite elements methods.

8. Masarik [MAS85] applies simulated annealing to the following problem: given N positive real numbers $m_1, \ldots, m_N$, find a permutation π of the set of integers $\{1, \ldots, N\}$, such that the function

$$C(\pi) = \left| \sum_{j=1}^{N} m_{\pi(j)} \exp\left\{2\pi i \frac{j}{N}\right\} \right| \tag{7.34}$$

is minimal. One of the applications of this problem is the minimization of the static unbalance of a stem turbine circular wheel (the m_i's correspond to the weights of the turbine paddles).

Chapter 8

Some miscellaneous topics

In this chapter, parallel implementations of the simulated annealing algorithm on *multi-processor architectures* (section 8.1) are described as well as an extension of the algorithm to problems with continuous variables (section 8.2).

8.1 Parallel implementations

As is clear from the discussion of applications in chapters 6 and 7, one of the major disadvantages of simulated annealing is that application of the algorithm may require large amounts of computation time (e.g., in the discussion of placement algorithms in subsection 7.2.2, CPU times of more than 84 hours on a VAX 11/780-computer are reported). *Hardware acceleration* of the simulated annealing algorithm can be achieved by rewriting the algorithm in microcode to execute it on fast *micro-engines* attached to a multi-processor host. Spyra and Hage [SPI85] report on a *micro-coded implementation* of the simulated annealing algorithm applied to the problem of gate-array placement achieving a speedup of a factor 20 compared to a high-level host implementation. These results are comparable to those obtained with implementations on large mainframes.
Besides hardware acceleration a substantial gain in computational effort can be obtained by parallelizing the algorithm in order to be able to execute it on a multi-processor architecture. However, such a parallelization is not a trivial task, since the simulated annealing al-

gorithm is intrinsically of a sequential nature: transitions are carried out one after the other.
The research on parallel implementations of the simulated annealing algorithm evolves very fastly over the last two years. Work is presented on generally applicable parallel simulated annealing algorithms and parallel algorithms tailored to a given problem instance. In this monograph we concentrate on a description of the general algorithms presented by Aarts *et al.* [AAR86a], [AAR86b] (subsection 8.1.1) and a discussion of a number of concepts used in various tailored algorithms related to applications in computer-aided circuit design (subsection 8.1.2). A few remarks on special parallel implementions are given in subsection 8.1.3. Due to the fast evolution of this subject we anticipate that much new material will be published between the time we write this monograph and the day it appears. However, most of the ideas covered here are straightforward approaches to the problem of parallelization of the annealing algorithm and can, therefore, serve as a basic introduction to the probably more sophisticated algorithms that may be proposed in the near future.

8.1.1 General algorithms

The parallel implementations described by Aarts *et al.* [AAR86a], [AAR86b] are based on the homogeneous algorithm (see chapter 3). Essential features of the first of the two approaches described in [AAR86b] are the assignment of a Markov chain to each of the available processors and the equal length of all Markov chains (see also [AAR86a]). The chains are executed in parallel and during the execution information is transferred from a given chain to its successor. This approach is designated as the *systolic* algorithm. In order to be able to let processors execute Markov chains simultaneously, each Markov chain is divided into a number of *sub-chains* equal to the number of available processors. Execution of Markov chain K is started as soon as execution of the *first* sub-chain of Markov chain $K-1$ is completed.
The exchange of information between processors is determined by the requirement that quasi-equilibrium should be preserved throughout execution of the algorithm. To explain this point, consider the exe-

cution of Markov chain K on processor i. The initial configuration and value of the control parameter are received from processor $i-1$, executing Markov chain $K-1$, after completion of the first sub-chain of Markov chain $K-1$. Next, a sub-chain is generated and after completion of this sub-chain, two sets of data are available to continue the execution of the Markov chain: the present configuration and value of c as well as the configuration and value of c (updated according to the decrement rule given by eq. 5.32) of processor $i-1$, after completion of its *second* sub-chain (see figure 8.1). A probabilistic choice

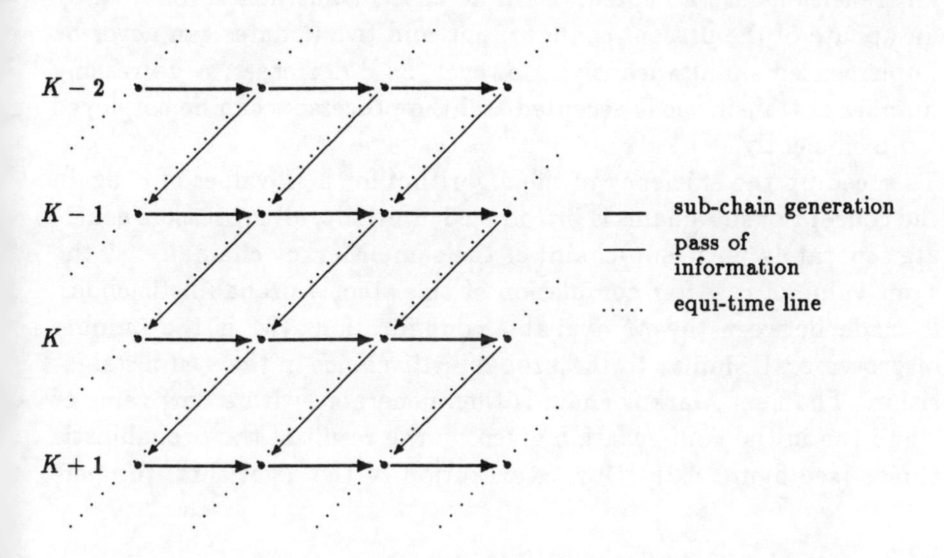

Figure 8.1: Schematic representation of the systolic algorithm (reproduced from [AAR86b]).

is made between these two sets. The probability of choosing one of the two possibilities is given by the normalized probability of occurrence of the configurations in the stationary distribution of eq. 3.40. It is assumed that the latter manipulation is a good way to preserve quasi-equilibrium, in spite of the fact that execution of a subsequent Markov chain is started before execution of the previous one is terminated.
The employed cooling schedule is the one discussed in section 5.3 (see

[AAR85a]). Note that in this implementation, all processors have about equal computational loads, since all Markov chains have the same length. Moreover, there is a relatively small amount of communication between processors.

In the second approach described by Aarts *et al.* [AAR86b], the *clustered* algorithm, all processors are used to generate cooperatively the same Markov chain. Taking into account the fact that accepted transitions can only be implemented consecutively, this implementation has a low efficiency for high values of c. For, if c is large, virtually all transitions are accepted. Each accepted transition is followed by an update of the present configuration and two updates can never be implemented simultaneously. However, as c decreases, a decreasing number of transitions is accepted and the processors can be employed more efficiently.
To speed up the efficiency of the algorithm for high values of c, again the concept of sub-chains is introduced. Initially, all processors generate separately part (sub-chain) of the same Markov chain, i.e. at the same value of c. After completion of this step, a probabilistic choice is made between the M available configurations (M is the number of processors), similar to the probabilistic choice in the systolic algorithm. The next Markov chain is then generated with a new value of c and the initial configuration given by the result of the probabilistic choice (see figure 8.2). During execution of the algorithm, the per-

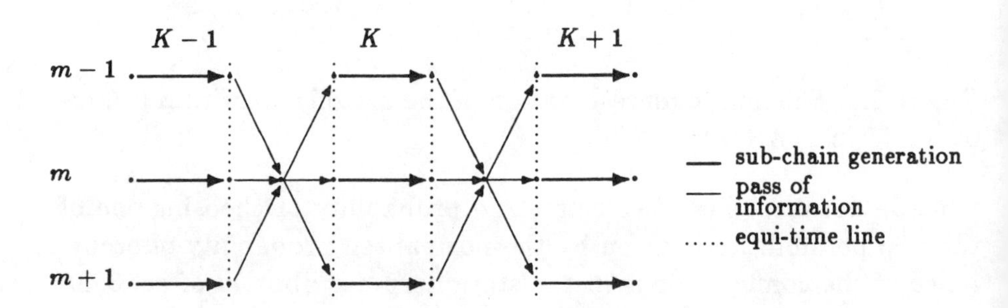

Figure 8.2: Schematic representation of the initial stages of the clustered algorithm (reproduced from [AAR86b]).

centage of accepted transitions is predicted that would exist if the processors were clustered two-by-two (each group of processors generating part of the same Markov chain). As soon as this predicted value exceeds 50%, the clustering is effected. Each group of processors then generates part of the same Markov chain, but this part is now twice as large as before. Each time the predicted percentage of accepted transitions for a new clustering (four-by-four, eight-by-eight etc.) exceeds 50%, clustering is repeated until finally all processors generate simultaneously the same Markov chain.
For both algorithms, polynomial-time execution is proved along the lines sketched in section 6.2 for the sequential algorithm.
Numerical results are obtained by solving a number of TSPs, for which the locations of the cities are positioned on a regular grid (see figure 4.1 and section 6.4). From these results it is concluded that the performance of the systolic algorithm decreases as the number of processors increases, which can be explained by the fact that for an increasing number of processors, the length of the sub-chains decreases and eventually becomes too small to account for the preservation of quasi-equilibrium. Consequently, the parameter δ (determining the decrement in c, given by eq. 5.32), has to be decreased to obtain the same quality of solution, which results in a larger number of Markov chains and decreasing efficiency. As the number of processors is further increased, this effect becomes more pronounced and for a given problem size a critical number of processors is eventually reached for which no gain in computational effort is achieved anymore.
For the clustered algorithm, the efficiency is almost 100% without loss in quality of solution, i.e. usage of M processors reduces total computation time by a factor M with the same quality of solution as the sequential algorithm. Theoretically, the efficiency should not deteriorate with increasing number of processors; however, it is argued that with a large number of processors, communication problems will limit the efficiency.

8.1.2 Tailored algorithms

The literature provides a vast number of papers that present parallelization strategies of the simulated annealing algorithm that are tailored to the problem under consideration. By tailored we mean that parallelization is obtained by partitioning the subtasks or data subsets as they typically exist for a given problem instance. In [FELT85], for example, Felten *et al.* consider the TSP and show that a configuration of processors given by a *hypercube* can be effectively exploited to solve TSPs via simulated annealing. The cities of a TSP are evenly distributed over the available processors. Two manipulations are combined: each processor determines for its present set of cities an optimal (sub-)tour via simulated annealing and occasionally cities are exchanged between processors (the possibility of exchange determined by the position of the processors on the hypercube). Efficiencies of almost 100% are reported.

Most of the tailored algorithms reported in the literature are constructed for VLSI layout problems, viz. (standard) cell placement (see subsection 7.2.2), taking the TimberWolf package [SEC85], [SEC86b] as a starting point (Banerjee and Jones [BAN86], Casotto *et al.* [CAS86], Kravitz and Rutenbar [KRA86a], [KRA86b], [RUT86], Rose *et al.* [ROSE86]), global routing (see subsection 7.4) (Chung and Rao [CHU86] and Darema-Rogers *et al.* [DAR87]) or topological array-logic minimization (Devadas and Newton [DEV86b]). These problems are well suited for a partitioning approach since they involve many complex calculations on an area map that can be easily divisioned. Furthermore, we mention that partitioning approaches have been applied earlier to parallel hardware machines using iterative improvement algorithms in placement problems (Chyan and Breuer [CHY83] and Ueda *et al.* [UED83]).

Casotto *et al.* [CAS86] achieve parallelization of the simulated annealing algorithm by partitioning the set of cells in a macro-cell placement problem into as many subsets as available processors and then assigning each subset to a different processor. The processors are allowed to run asynchronously as they independently calculate transitions among configurations, assuming that all cells assigned to other processors are fixed, thus introducing errors in the calculation of the cost function. It is argued that these errors disappear as the value of the control parameter becomes small and that they do not effect the global performance of the algorithm. Furthermore, the authors identify a clustering problem which is defined as: find an assignment of clusters of

cells to processors such that the cells in a cluster occupy neighbouring positions on the chip. Since the positioning of the cells changes during the execution of the algorithm a static partitioning is not satisfactory. Therefore, an *auxiliary* simulated annealing algorithm is used to calculate the appropriate partitioning at the beginning of each Markov chain. The auxiliary algorithm runs together with the main algorithm exactly at the same value of the control parameter. It is concluded that the effect of the clustering is not critical with respect to the performance of the algorithm.
The problem of erroneously calculated cost differences in parallel annealing implementations for global routing is addressed by Darema-Rogers *et al.* [DAR87]. They conclude that parallel algorithms allowing these errors during the generation process of a Markov chain combined with restoration of the correct value of the cost function at the end of each Markov chain yields an efficient implementation.
Other methods using spatial partitioning techniques are presented by Banerjee and Jones [BAN86] and Devadas and Newton [DEV86b]. The latter method is applied to topological optimization of multiple array logic and uses a *dynamical windowing technique* to generate time-dependent allocations of spatial regimes of a logic array to each available processor.

The simulated annealing algorithm exhibits markedly different characteristics in different stages of the algorithm (see e.g. section 4.3). At large values of the control parameter the acceptance ratio is large (close to 1), whereas at small values the acceptance ratio is typically very small. In their approaches to the problem of parallelization of the simulated annealing algorithm Rose *et al.* [ROSE86] as well as Kravitz and Rutenbar [KRA86a], [KRA86b], [RUT86] introduce adaptive strategies based on the use of different algorithms in the different regimes of the control parameter. Rose *et al.* first apply a randomization algorithm based on a repeated iterative improvement strategy using different starting configurations. Clearly this can be done in a fully parallel way. Next they apply a parallel simulated annealing algorithm running at small values of the control parameter. Here again they apply a spatial partitioning algorithm in which the placement area is divided over the available processors. Errors in the calculation of the cost function are allowed. Similar to Darema-Rogers *et al.* [DAR87], they include a means of reducing the amount of errors. More specifically, they use a sophisticated broadcasting strategy instead of repeated recalculation of the correct value of the cost function as in [DAR87]. Kravitz and Rutenbar use different parallel simulated annealing algorithms in the two different regimes of the value of the control parameter. In the large-value regime they apply a *single move decomposition algorithm* which simply divides the task of computing a single transition into several subtasks that can be evaluated in parallel. In their placement application this is a sensible approach since each transition takes a substantial computational effort. In the small-value regime they apply a *parallel moves algorithm* which uses the concept of *serializable subsets* of transitions, i.e. a set of non-interacting transitions. Determination of serializable subsets for a given problem instance is a difficult problem and needs to be repeated often. Therefore, they adopt the idea that serializable subsets can be obtained by rejecting accepted transitions, i.e. only the first accepted transition found during the parallel evaluation of transitions is counted, all the other accepted transitions are aborted. Thus if N transitions are calculated out of which M are accepted, then $N - M + 1$ transitions are counted as trials in the generation of a Markov chain. The speed up of the single move decomposition algorithm is found to be constant (≈ 2) as a function of the control parameter. The efficiency of the parallel moves algorithm is small (≈ 1) in the large-value regime and large (≈ 3.5) in the small-value regime of the control parameter (the numbers refer to an implementation on a 4-processor computer system). Thus at some value of the control parameter it becomes favorable to switch from the move decomposition to the parallel moves algorithm. The switching value is determined during execution of the algorithm using a probabilistic model. We mention that the parallel moves algorithm by nature is a general algorithm since it is not tailored to the structure of a given problem instance.

For most of the algorithms treated in this subsection efficiencies are reported up to 80% refering to implementations on multi-processor systems with upto 8 processors. The algorithm proposed by Banerjee and Jones [BAN86] was run on a GOULD 9050 computer system simulating a 6-dimensional hypercube (64 processors). Optimistic estimates indicate that implementation of the algorithm on an AMETEK hypercube processor system should be at least an order of magnitude faster than the TimberWolf package run on a VAX11/780.

8.1.3 Special parallel implementations

Special parallel implementations of the simulated annealing algorithm on dedicated hardware machines are presented by several authors (Aarts and Korst [AAR87c], Černy and Novák [ČER83], [ČER86a] and Chung and Rao [CHU86]). Characteristic for these implementations is the fact that massive parallelism is used and that they show a resemblance with neural networks [HOP82], [HOP85]. In [CHU86], Chung and Rao present models for hardware partitioning and routing engines. Massive parallelism is achieved by mapping the problem onto an array of processors, each processor calculating the effect of a local rearrangement in a global configuration. In [AAR87c], Aarts and Korst propose a model to solve combinatorial optimization problems on massively parallel computer architectures based on the Boltzmann machine (see subsection 7.3.2). For two distinct 0-1 integer formulations of the travelling salesman problem (as an *assignment problem* and as a *quadratic assignment problem*) it is shown that near-optimal solutions can be obtained by mapping the corresponding 0-1 variables onto discrete computing elements and transforming the cost functions into the consensus function associated with the Boltzmann machine. Results of computer simulations are presented for two instances of the travelling salesman problem (i.e. with 10 and 30 cities).
Application of simulated annealing to image processing (see subsection 7.3.1) is very well suited for massively parallel approaches. Geman and Geman [GEM84], for example, introduce a parallel simulated annealing algorithm for restoring binary images based on *stochastic relaxation* techniques. Their algorithm can be easily mapped onto a massively parallel pixel-oriented multi-processor system. Parallelism is achieved by assigning each pixel in an image and its immediate "neighbourhood" to a processor. This can be done since transitions in the stochastic relaxation algorithm are based upon the current value of a pixel and the values of the neighbouring pixels which are

sampled from a local conditional probability distribution. A similar idea is used by Černy and Novák in [ČER86a], where they propose a network of ***stochastically coupled processors*** [ČER83] to solve the problem of object identification in a noisy scene. We are aware of many places where research is presently carried out to combine simulated annealing and image processing on massively parallel computer systems and we anticipate that this subject may play an important role in the near future.

8.2 Continuous optimization

8.2.1 Generalization of simulated annealing to continuous problems

In [VAND84], Vanderbilt and Louie describe an extension of the simulated annealing algorithm to problems with continuous variables. Again a function C is given, but this time C is defined over an n-dimensional *continuous* variable space, i.e.

$$C(\mathbf{x}) = C(x_1, \ldots, x_n), \tag{8.1}$$

where $x_i \in \mathbf{R}$, $i = 1, \ldots, n$. The problem is to find $\mathbf{x}_{opt}$, satisfying

$$C(\mathbf{x}_{opt}) = \min\{C(\mathbf{x}) \mid \mathbf{x} \in \mathbf{R}^n\}. \tag{8.2}$$

In order to be able to apply simulated annealing, a generation mechanism for transitions is to be defined, i.e. given the present configuration $\mathbf{x}$, a way to perturb $\mathbf{x}$ into a new configuration $\mathbf{y}$. The probability that $\mathbf{y}$ will be accepted is again given by the Metropolis criterion:

$$Pr\{\hat{\mathbf{X}}(k) = \mathbf{y} \mid \hat{\mathbf{X}}(k-1) = \mathbf{x}\} = \min\{1, \exp\left(-\frac{C(\mathbf{y}) - C(\mathbf{x})}{c}\right)\}. \tag{8.3}$$

Again, it can be shown that

$$\lim_{k \to \infty} Pr\{\hat{\mathbf{X}}(k) = \mathbf{x}\} = \frac{1}{Z} \exp\left(-\frac{C(\mathbf{x})}{c}\right), \tag{8.4}$$

where Z is a normalizing factor. $Pr\{\hat{\mathbf{X}}(k) = \mathbf{x}\} \cdot d^n x$ now denotes the probability for $\hat{\mathbf{X}}(k)$ to be in a volume $d^n x$.
Vanderbilt and Louie propose the following choice for the generation mechanism for transitions, i.e. the choice of $\Delta\mathbf{x}$, defining $\mathbf{y} = \mathbf{x} + \Delta\mathbf{x}$. First, a vector $\mathbf{u}$ is constructed, whose n components are randomly drawn from a uniform distribution on $[-\sqrt{3}, \sqrt{3}]$, i.e. the u_i have mean 0 and variance 1. Next, $\Delta\mathbf{x}$ is chosen according to

$$\Delta\mathbf{x} = Q \cdot \mathbf{u}, \tag{8.5}$$

where the matrix Q is used to control the choice of $\Delta\mathbf{x}$. Vanderbilt and Louie show that the *covariance matrix* S, defined as

$$S_{ij} = \int (\Delta x)_i (\Delta x)_j g(\mathbf{u}) d^n u, \tag{8.6}$$

where $g(\mathbf{u})$ is the probability density function of the vector $\mathbf{u}$, is given by

$$S = Q \cdot Q^T. \tag{8.7}$$

Thus, if the covariance matrix is given, $\Delta\mathbf{x}$ is obtained by first solving eq. 8.7 for Q and then using eq. 8.5. Thus, the problem is shifted from determining $\Delta\mathbf{x}$ to determining the covariance matrix S.
The matrix S should in some way reflect the topology of the search space. It is well known, for example, that in the neighbourhood of a minimum, where C can be approximated by a quadratic function, S should be proportional to the inverse Hessian, i.e.

$$S^{-1} \propto (\nabla^2 C(\mathbf{x}))_{\mathbf{x}=\mathbf{x}_{opt}}. \tag{8.8}$$

Vanderbilt and Louie propose the use of the values of $\mathbf{x}(k)$ (the configurations of the Markov chain) as a measure of the local topology of the search space, i.e. for a given Markov chain of length L, the following two quantities are calculated:

$$X_i^{(1)} = \frac{1}{L} \sum_{k=1}^{L} x_i(k) \tag{8.9}$$

and

$$X_{ij}^{(2)} = \frac{1}{L} \sum_{k=1}^{L} (x_i(k) - X_i^{(1)})(x_j(k) - X_j^{(1)}) \tag{8.10}$$

and the matrix S for the new Markov chain is now chosen proportional to $X^{(2)}$.
Results are presented for a number of well-known problems from the literature, from which it is concluded that the proposed algorithm is competitive with existing optimization algorithms, in terms of both quality of results and computation time.

Similar approaches are proposed by Černy [ČER84], Wille and Vennik [WIL85], Khachaturyan [KHA86] and Bohachevsky *et al.* [BOH86a], [BOH86b], with different (and less sophisticated) choices for the search direction. In Černy's approach [ČER84], the search direction ($\Delta \mathbf{x}$) is randomly drawn from the set of axes of the coordinate system. After each move the coordinate system is transformed in such a way that the first axis always points towards the best configuration currently found. The next configuration is set to $\mathbf{x} + \rho \cdot \Delta \mathbf{x}$, where ρ is a number drawn from a probability distribution, whose parameters are adjusted if the acceptance ratio drops below or exceeds certain critical values. In the approach of Bohachevsky *et al.* [BOH86a] and [BOH86b], $\Delta \mathbf{x}$ is set to $q \cdot \mathbf{u}$, where q is a (predefined) step size and $\mathbf{u}$ a vector with randomly drawn components in the interval $[-1, 1]$. An interesting aspect of the approach of Bohachevsky *et al.* is that instead of the usual acceptance criterion, given by eq. 8.3, they propose to use

$$Pr\{\hat{\mathbf{X}}(k) = \mathbf{y} \mid \hat{\mathbf{X}}(k-1) = \mathbf{x}\} = \min\{1, \exp\left(-\frac{C(\mathbf{y}) - C(\mathbf{x})}{c \cdot (C(\mathbf{y}) - \tilde{C}_{min})}\right)\}, \tag{8.11}$$

where $\tilde{C}_{min}$ is an estimate for the minimum value of the objective function C. The algorithm is run for a fixed value of c and by introducing the term $(C(\mathbf{y}) - \tilde{C}_{min})$ in the denominator of eq. 8.11 it is ensured that the probability to accept deteriorations tends to zero as the algorithm approaches a global minimum. It remains an open question, however, whether a global minimum is indeed approached. Applications to optimal design problems are discussed.
Finally, Szu and Hartley [SZU87] claim that for continuous optimization a decrement rule of the form

$$c_{k+1} = (l + c_k)^{-1}, \tag{8.12}$$

where l is a real-valued constant, is sufficient to establish convergence to a global minimum, provided Δx is generated according to a *Cauchy distribution*. Note, that eq. 8.12 allows a faster decrement of c_k than the logarithmic expression given by eq. 3.105.

8.2.2 Applications of simulated annealing to continuous problems

Finally, some applications of simulated annealing to continuous optimization problems are briefly described:

1. Wille [WIL85], [WIL86b], [WIL87a] uses simulated annealing, for example, for the determination of *minimum-energy configurations* of N atoms, interacting under two-body Lennard-Jones forces. For $N \leq 25$, all previously known best results are also found by simulated annealing, for $N = 24$ a new minimum better than any previous result is reported.

2. In [BIS86], Biswas and Hamann apply simulated annealing to obtain minimum-energy configurations of *silicon clusters*. Their model is based on *Langevin molecular dynamics*, where the system of particles is connected to a heat bath that provides random stochastic forces as well as viscous friction. The authors report very satisfactory results.

3. Semenovskaya *et al.* [SEM85] use simulated annealing to determine *crystal structures* from the low-temperature equilibrium state of a non-ideal gas model composed of atoms within a crystal unit cell. Their approach is based on the assumption that the crystal structure is determined by the spatial atomic distribution in a unit cell that minimizes the difference between the calculated and experimentally observed X-ray diffraction reflexion intensities.

4. In [NIC84], Nicholson *et al.* use simulated annealing to obtain pairs of *distribution functions* in coordinate and wave vector space, that are suitable for realistic electronic structure calculations in *disordered metals*. The objective in this application is

to find pair distribution functions that minimize the difference with experimentally determined functions that satisfy a set of constraints, while being interrelated by a Fourier transformation.

Chapter 9

Summary and conclusions

In this monograph we have presented a review of the theory and applications of simulated annealing. The simulated annealing algorithm is formulated using the theory of Markov chains and the convergence of the algorithm is analysed within this scope.

The asymptotic convergence of the algorithm is formulated in terms of a homogeneous algorithm (generation of a sequence of Markov chains) and an inhomogeneous algorithm (generation of a single inhomogeneous Markov chain). For the homogeneous algorithm convergence to a globally minimal configuration has been proved by numerous authors along the same lines, the only difference in the respective proofs being the conditions imposed on the transition matrix of the Markov chains. For the inhomogeneous algorithm, several authors have shown that convergence to global minima is obtained if, along with conditions on the transition matrix, certain conditions are imposed on the sequence of values of the control parameter. The strongest of these conditions is both necessary and sufficient.

Since the aforementioned convergence results are of an asymptotic nature, any implementation of the algorithm requires approximate values for certain parameters governing the convergence of the algorithm. A number of such sets of parameter choices or cooling schedules are discussed (all schedules are based on approximations for the homogeneous algorithm; we are not aware of cooling schedules in which the value of the control parameter is decreased after each transition (inhomogeneous algorithm)). The schedules presented are classified into two categories: those for which the Markov chain length increases and the decrement of the control parameter is kept constant throughout execution of the algorithm, and those for which the Markov chain length is kept constant and the decrement is made dependent on the evolution of the algorithm. Moreover, this classification more or less coincides with a distinction between conceptually simple and more elaborate schedules.

From the relation between statistical physics and simulated annealing it is concluded that macroscopic ensemble averages can be usefully employed in the analysis of the behaviour of the simulated annealing algorithm. The relation between spin-glass Hamiltonians and cost functions for some optimization problems raises a number of interesting questions, e.g. with respect to the existence of an ultrametric structure of the configuration space, the cooling rate dependence and the existence of phase transitions.

The performance of the simulated annealing algorithm is discussed in terms of both running time and quality of the final solution obtained by the algorithm. Theoretical worst-case running time results are presented for some cooling schedules, i.e. it is shown that some cooling schedules lead to a polynomial-time execution of the algorithm. Theoretical worst-case results for the quality of the solution obtained by the algorithm are only available for one particular problem, but not in general (see the open problems at the end of this chapter).
Many empirical average-case results have been obtained for both running time and quality of the final solution by solving large numbers of instances of several well-known combinatorial optimization problems, as for example the travelling salesman problem, the quadratic assignment problem, the graph partitioning problem and the graph colouring problem. An important conclusion from these results is that the performance of the algorithm is strongly dependent on the chosen cooling schedule, especially as far as the quality of solution is concerned. Indeed, it is shown that the performance of the algorithm deteriorates severely if the cooling schedule employed belongs to the class of simple schedules mentioned before.
Some extensive computational experiments are discussed, in which the

simulated annealing algorithm is pitted against tailored heuristics for some of the aforementioned problems. Unfortunately, in these experiments the cooling schedules employed belong to the class of simple schedules; still, the algorithm is shown to be competitive (in terms of quality of solution) with the tailored heuristics, although in most cases at the cost of large amounts of computation time. Furthermore, as Skiscim and Golden [SKI83] already state: "It is perhaps unfair to expect a new approach (...) to compete with the best of an almost endless array of TSP heuristics without extensive tuning". For the class of graph partitioning problems the quality of solutions obtained by the simulated annealing algorithm is substantially better than those obtained by a tailored heuristic, using computation times of the same order of magnitude.

Many authors claim the simulated annealing algorithm to be a widely applicable optimization technique and the rich variety of applications discussed in this monograph strongly supports this claim; applications are presented in such diverse areas as computer-aided design of integrated circuits, image processing, neural network theory etc. In order to use the algorithm to solve a particular problem, a number of items have to be defined: a set of configurations, a generation mechanism for transitions and a cost function. For some problems it is hard to define the first and the last item (see e.g. the channel routing problem described in subsection 7.2.3) and some skill is required from the user; once a set of configurations is defined, the definition of a generation mechanism is usually straightforward.

As far as the performance of the algorithm is concerned, it can be concluded that for many applications the quality of the solution obtained by the algorithm is at least as good as and sometimes much better than the one obtained by previous algorithms, if they exist at all. The latter seems to be especially true for randomly generated problems (cf. the results for large, randomly generated, combinatorial optimization problems in section 6.3 and some of the problems discussed in chapter 7). However, a considerable price has to be paid in terms of long computation times: CPU times of more than 84 hours on a VAX 11/780-computer are reported for a large placement problem [SEC85] (on the other hand, the solution then obtained is 66%

better than the one obtained by a previous algorithm). It is shown that parallel execution of the algorithm might drastically reduce these computation times.

Finally, we formulate some open problems and remaining questions:

- it is desirable to make a profound comparison between different implementations of the simulated annealing algorithm, i.e. with different cooling schedules, and to compare the best implementation (i.e. the best cooling schedule) with tailored heuristics for a large set of combinatorial optimization problems;

- given a particular instance of a combinatorial optimization problem and a particular implementation of the algorithm, it would be of great help to have a probabilistic measure for the deviation of the solution, obtained by the algorithm, from a globally minimal configuration.

As a final conclusion we state that the simulated annealing algorithm is a generally applicable, flexible, robust and easy-to-implement approximation algorithm, that is able to obtain near-optimal solutions for a wide range of optimization problems. However, computation times can be long and in a number of cases valuable tailored algorithms are available that can be executed in far less computation time. For those problem areas where no tailored algorithms are available, we consider simulated annealing to be a powerful optimization tool.

Bibliography

[AAR84] Aarts, E.H.L, P.J.M. van Laarhoven, R. Burgess and B. Culleton, Linear Arrangement using Statistical Cooling, *Philips Research Report*, 1984.

[AAR85a] Aarts, E.H.L and P.J.M. van Laarhoven, Statistical Cooling: A General Approach to Combinatorial Optimization Problems, *Philips J. of Research*, 40(1985)193-226.

[AAR85b] Aarts, E.H.L and P.J.M. van Laarhoven, A New Polynomial Time Cooling Schedule, *Proc. IEEE Int. Conference on Computer-Aided Design*, Santa Clara, November 1985, pp. 206-208.

[AAR86a] Aarts, E.H.L., F.M.J. de Bont, J.H.A. Habers and P.J.M. van Laarhoven, A Parallel Statistical Cooling Algorithm, *Proc. STACS 86*, *Springer Lecture Notes in Computer Science*, 210(1986)87-97.

[AAR86b] Aarts, E.H.L., F.M.J. de Bont, J.H.A. Habers and P.J.M. van Laarhoven, Parallel Implementations of the Statistical Cooling Algorithm, *Integration*, 4(1986)209-238.

[AAR86c] Aarts, E.H.L. and J.H.M. Korst, Simulations of Learning in Parallel Networks Based on the Boltzmann Machine, *Proc. 2nd European Simulation Congress*, Antwerp, September 1986, pp. 391-398.

[AAR87a] Aarts, E.H.L. and P.J.M. van Laarhoven, Simulated Annealing: A Pedestrian Review of the Theory and Some Applications, in: P.A. Devijver and J. Kittler, eds., *Pattern Recognition, Theory and Applications*, (Springer Verlag, Berlin, 1987), pp. 179-192.

[AAR87b] Aarts, E.H.L. and P.J.M. van Laarhoven, A Pedestrian Review of the Theory and Application of the Simulated Annealing Algorithm, in: J.L. van Hemmen and I. Morgenstern, eds.,

Heidelberg Colloquium on Glassy Dynamics and Optimization, *Springer Lecture Notes in Physics*, 275(1987)288-306.

[AAR87c] Aarts, E.H.L. and J.H.M. Korst, Boltzmann Machines and Their Applications, *Proc. PARLE 87*, *Springer Lecture Notes in Computer Science*, 258(1987)34-50.

[AAR88a] Aarts, E.H.L., J.H.M. Korst and P.J.M. van Laarhoven, A Quantitative Analysis of the Simulated Annealing Algorithm: A Case Study for the Traveling Salesman Problem, *J. Statis. Phys.*, 50(1988)189-206.

[AAR88b] Aarts, E.H.L. and J.H.M. Korst, *Simulated Annealing and Boltzmann Machines*, (Wiley, Chichester, 1988).

[AAR89] Aarts, E.H.L. and P.J.M. van Laarhoven, Simulated Annealing: An Introduction, to appear in: *Statistica Neerlandica*, 43(1989).

[ACK85] Ackley, D.H., G.E. Hinton and T.J. Sejnowski, A Learning Algorithm for Boltzmann Machines, *Cognitive Science*, 9(1985)147-169.

[ANI87a] Anily, S. and A. Federgruen, Ergodicity in Parametric Nonstationary Markov Chains: An Application to Simulated Annealing Methods, *Oper. Res.*, 35(1987)867-874.

[ANI87b] Anily, S. and A. Federgruen, Simulated Annealing Methods with General Acceptance Probabilities, *J. Appl. Prob.*, 24(1987)657-667.

[ARM84] Armstrong, B.A., A Hybrid Algorithm for Reducing Matrix Bandwidth, *Int. J. for Num. Methods in Eng.*, 20(1984)1929-1940.

[BAN86] Banerjee, P. and M. Jones, A Parallel Simulated Annealing Algorithm for Standard Cell Placement on a Hypercube Computer, *Proc. IEEE Int. Conference on Computer-Aided Design*, Santa Clara, November 1986, pp. 34-37.

[BAR76] Barker, J.A. and D. Henderson, What is "liquid"? Understanding the states of matter, *Reviews of Modern Physics*, 48-4(1976)587-671.

[BAU86] Baum, E.B., Iterated Descent: A Better Algorithm for Local Search In Combinatorial Optimization Problems, Cal. Institute of Technology, Pasadena, *unpublished manuscript*, 1986.

[BEA59] Beardwood, J., J.H. Halton and J.M. Hammersley, The Shortest Path Through Many Points, *Proc. Cambridge Philos. Soc.*, 55(1959)299-327.

[BEE85] Beenker, G.F.M., T.A.C.M. Claasen and P.W.C. Hermens, Binary Sequences with Maximally Flat Amplitude Spectrum, *Philips J. of Research*, 40(1985)289-304; Corrigendum, 40(1985)399.

[BER87] Bernasconi, J., Low Autocorrelation Binary Sequences: Statistical Mechanics and Configuration Space Analysis, *J. Physique*, 48(1987)559-567.

[BIN78] Binder, K., *Monte Carlo Methods in Statistical Physics*, (Springer, New York, 1978).

[BIS86] Biswas, R. and D.R. Hamann, Simulated annealing of silicon atom clusters in Langevin molecular dynamics, *Phys. Rev. B*, 34(1986)895-901.

[BOH86a] Bohachevsky, I., V.K. Viswanathan and D.R. Rossbach, Generalized Simulated Annealing in the Construction of "Intelligent" Design Programs, *Workshop on Statistical Physics in Engineering and Biology*, Lech, July 1986.

[BOH86b] Bohachevsky, I., M.E. Johnson and M.L. Stein, Generalized Simulated Annealing for Function Optimization, *Technometrics*, 28(1986)209-217.

[BON84] Bonomi, E. and J.-L. Lutton, The N-city Travelling Salesman Problem: Statistical Mechanics and the Metropolis Algorithm, *SIAM Rev.*, 26(1984)551-568.

[BON86] Bonomi, E. and J.-L. Lutton, The Asymptotic Behaviour of Quadratic Sum Assignment Problems: A Statistical Mechanics Approach, *European J. of Oper. Res.*, 26(1986)295-300.

[BOU86] Bounds, D.G., Physics for Travelling Salesmen: Some New Approaches to Combinatorial Optimization, submitted to the *Bulletin of the Institute of Mathematics and its Applications*, 1986.

[BRAU86] Braun, D., C. Sechen and A.L. Sangiovanni-Vincentelli, Thunderbird: A Complete Standard Cell Layout System, *Proc. 1986 Custom IC Conf.*, Rochester, May 1986, pp. 276-280.

[BRAY84] Brayton, R.K., G.D. Hachtel, C.T. McMullen and A.L. Sangiovanni-Vincentelli, *Logic Minimization Algorithms for VLSI Synthesis*, (Kluwer Academic Publishers, Boston, 1984).

[BRE72] Breuer, M.A., ed., *Design Automation of Digital Systems, vol. 1: Theory and Techniques*, (Prentice-Hall, Englewood Cliffs, 1972).

[BUR83a] Burkard, R.E. and U. Fincke, The asymptotic probabilistic behaviour of quadratic sum assignment problems, *Z. Operations Research*, 27(1983)73-81.

[BUR83b] Burkard, R.E. and T. Bönninger, A heuristic for quadratic boolean programs with applications to quadratic assignment problems, *European J. of Oper. Res.*, 13(1983)374-386.

[BUR84] Burkard, R.E. and F. Rendl, A thermodynamically motivated simulation procedure for combinatorial optimization problems, *European J. of Oper. Res.*, 17(1984)169-174.

[CARN85] Carnevalli, P., L. Coletti and S. Paternello, Image processing by simulated annealing, *IBM J. Res. Develop.*, 29(1985)569-579.

[CART86] Carter, H.W., Computer-Aided Design for Integrated Circuits, *Computer*, 19(1986)19-36.

[CAS86] Casotto, A., F. Romeo and A.L. Sangiovanni-Vincentelli, A Parallel Simulated Annealing Algorithm for the Placement of Macro-Cells, *Proc. IEEE Int. Conference on Computer-Aided Design*, Santa Clara, November 1986, pp. 30-33.

[CAT85] Catthoor, F., H. DeMan and J. Vanderwalle, SAILPLANE: A Simulated Annealing based CAD-tool for the Analysis of Limit-Cycle Behaviour, *Proc. IEEE Int. Conference on Computer Design*, Port Chester, October 1985, pp. 244-247.

[CAT86] Catthoor, F., H. DeMan and J. Vanderwalle, Investigation of Finite Word-length effects in Arbitrary Digital Filters using Simulated Annealing, *Proc. IEEE Int. Symposium on Circuits and Systems*, San Jose, May 1986, pp. 1296-1297.

[ČER83] Černy, V., Multiprocessor System as a Statistical Ensemble: A Way Towards General-Purpose Parallel Processing and MIMD Computers?, Comenius University, Bratislava, *unpublished manuscript*, 1983.

[ČER84] Černy, V., Minimization of Continuous Functions by Simulated Annealing, Research Institute for Theoretical Physics, University of Helsinki, *Preprint No. HU-TFT-84-51*, 1984.

[ČER85] Černy, V., Thermodynamical Approach to the Traveling Salesman Problem: An Efficient Simulation Algorithm, *J. Opt. Theory Appl.*, 45(1985)41-51.

[ČER86a] Černy, V. and I. Novák, Picture Processing by Statistically Coupled Processors: Relaxation Syntactical Analysis, *Workshop on Statistical Physics in Engineering and Biology*, Lech, July 1986.

[ČER86b] Černy, V., Solving Inverse Problems by Simulated Annealing, *Workshop on Statistical Physics in Engineering and Biology*, Lech, July 1986.

[CHU86] Chung, M.J. and K.K. Rao, Parallel Simulated Annealing for Partitioning and Routing, *Proc. IEEE Int. Conference on Computer Design*, Port Chester, October 1986, pp. 238-242.

[CHY83] Chyan, D.J. and M.A. Breuer, A Placement Algorithm for Array Processors, *Proc. 20th Des. Automation Conf.*, Miami Beach, June 1983, pp. 182-188.

[COL87] Collins, N.E., R.W. Eglese and B.L. Golden, Simulated Annealing - An Annotated Bibliography, Cambridge University, *preprint*, 1987.

[COM86] *Computer*, 19(1986).

[COO83] Cook, S.A., An Overview of Computational Complexity, *Comm. ACM*, 26(1983)400-408.

[CRO80] Crowder, H. and M.W. Padberg, Solving Large-Scale Symmetric Travelling Salesman Problems to Optimality, *Management Sci.*, 26(1980)495-509.

[DAN63] Dantzig, G.B., *Linear Programming and Extensions*, (Princeton University Press, Princeton, 1963).

[DAR87] Darema-Rogers, F., S. Kirkpatrick and V.A. Norton, Parallel Algorithms for Chip Placement by Simulated Annealing, *IBM J. Res. Develop.*, 31(1987)391-402.

[DEU76] Deutsch, D.N., A 'DOGLEG' Channel Router, *Proc. 13th Des. Automation Conf.*, San Francisco, June 1976, pp. 425-433.

[DEV86a] Devadas, S. and A.R. Newton, GENIE: A Generalized Array Optimizer for VLSI Synthesis, *Proc. 23rd Des. Automation Conf.*, Las Vegas, June 1986, pp. 631-637.

[DEV86b] Devadas, S. and A.R. Newton, Topological Optimization of Multiple Level Array Logic: On Uni and Multi-processors, *Proc. IEEE Int. Conference on Computer-Aided Design*, Santa Clara, November 1986, pp. 38-41.

[DEM83] De Micheli, G. and A.L. Sangiovanni-Vincentelli, PLEASURE: A Computer Program for Simple/Multiple Constrained/Unconstrained Folding of Programmable Logic Arrays, *Proc. 20th Des. Automation Conf.*, Miami Beach, June 1983, pp. 530-537.

[DIS86] Distante, F. and V. Piuri, Optimum Behavioral Test Procedure for VLSI Devices: A Simulated Annealing Approach, *Proc. IEEE Int. Conference on Computer Design*, Port Chester, October 1986, pp. 31-35.

[DUN85] Dunlop, A.E. and B.W. Kernighan, A Procedure for Placement of Standard-Cell VLSI Circuits, *IEEE Trans. on Computer-Aided Design*, CAD-4(1985)92-98.

[ELG87] El Gamal, A., L.A. Hemachandra, I. Shperling and V.K. Wei, Using Simulated Annealing to Design Good Codes, *IEEE Trans. Inform. Theory*, IT-33(1987)116-123.

[ETT85] Ettelaie, R. and M.A. Moore, Residual entropy and simulated annealing, *J. Physique Lett.*, 46(1985)L-893 - L-900.

[FAH83] Fahlman, S.E., G.E. Hinton and T.J. Sejnowski, Massively Parallel Architectures for AI: NETL, Thistle and Boltzmann Machines, *Proc. Nat. Conf. on AI, AAAI-83*, Washington D.C., August 1983, pp. 109-113.

[FELD85] Feldman, J.A. and D.H. Ballard, Connectionist Models and Their Properties, *Cognitive Science*, 9(1985)205-254.

[FELL50] Feller, W., *An Introduction to Probability Theory and Its Applications, vol. 1*, (Wiley, New York, 1950).

[FELT85] Felten, E., S. Karlin and S.W. Otto, The Traveling Salesman Problem on a Hypercubic, MIMD Computer, *Proc. 1985 Int. Conference on Parallel Processing*, St. Charles, August 1985, pp. 6-10.

[FIC82] Fiduccia, C.M. and R.M. Mattheyses, A Linear-Time Heuristic for Improving Network Partitions, *Proc. 19th Design Automation Conference*, Las Vegas, June 1982, pp. 175-181.

[FLE84] Fleisher, H., M.A. Tavel and D.B. Martin, Decomposition of Logic Functions by Simulated Annealing, IBM Corporation, Poughkeepsie, *unpublished manuscript*, 1984.

[FLE85] Fleisher, H., J. Giraldi, D.B. Martin, R.L. Phoenix and M.A. Tavel, Simulated Annealing as a Tool for Logic Optimization in a CAD Environment, *Proc. IEEE Int. Conference on Computer-Aided Design*, Santa Clara, November 1985, pp. 203-205.

[FU86] Fu, Y. and P.W. Anderson, Application of statistical mechanics to NP-complete problems in combinatorial optimization, *J. Phys. A: Math. Gen.*, 19(1986)1605-1620.

[GAR76] Garey, M.R., D.S. Johnson and L. Stockmeyer, Some Simplified NP-complete Graph Problems, *Theor. Comput. Sci.*, 1(1976)237-267.

[GAR79] Garey, M.R. and D.S. Johnson, *Computers and Intractability: A Guide to the Theory of NP-Completeness*, (W.H. Freeman and Co., San Francisco, 1979).

[GEL85] Gelfand, S.B. and S.K. Mitter, Analysis of Simulated Annealing for Optimization, *Proc. 24th Conf. on Decision and Control*, Ft. Lauderdale, December 1985, pp. 779-786.

[GEM84] Geman, S. and D. Geman, Stochastic Relaxation, Gibbs Distributions, and the Bayesian Restoration of Images, *IEEE Proc. Pattern Analysis and Machine Intelligence*, PAMI-6(1984)721-741.

[GIB02] Gibbs, J.W., *Elementary Principles in Statistical Mechanics*, (Yale University Press, New Haven, 1902).

[GID85a] Gidas, B., Nonstationary Markov Chains and Convergence of the Annealing Algorithm, *J. Statis. Phys.*, 39(1985)73-131.

[GID85b] Gidas, B., Global Optimization via the Langevin Equation, *Proc. 24th Conf. on Decision and Control*, Ft. Lauderdale, December 1985, pp. 774-778.

[GID86] Gidas, B., A Renormalization Group Approach to Image Processing Problems, to appear in: *IEEE Proc. Pattern Analysis and Machine Intelligence*, 1986.

[GOLA82] Golay, M.J.E., The Merit Factor of Long Low Autocorrelation Binary Sequences, *IEEE Trans. Inform. Theory*, IT-28(1982)543-549.

[GOLDB83] Goldberg, M.K. and M. Burstein, Heuristic Improvement Technique for Bisection of VLSI Networks, *Proc. IEEE Int. Conference on Computer Design*, Port Chester, November 1983, pp. 122-125.

[GOLDE80] Golden, B.L., L.D. Bodin, T. Doyle and W. Stewart Jr., Approximate Traveling Salesman Algorithms, *Oper. Res.*, 28(1980)694-711.

[GOLDE85] Golden, B.L. and W.R. Stewart, Empirical Analysis of Heuristics, in: E.L. Lawler, J.K. Lenstra, A.H.G. Rinnooy Kan and D.B. Shmoys, eds., *The Traveling Salesman Problem*, (Wiley, Chichester, 1985), pp. 207-249.

[GOLDE86] Golden, B.L. and C.C. Skiscim, Using Simulated Annealing to Solve Routing and Location Problems, *Naval Logistics Research Quarterly*, 33(1986)261-279.

[GON86] Gonsalves, G., Logic Synthesis Using Simulated Annealing, *Proc. IEEE Int. Conference on Computer Design*, Port Chester, October 1986, pp. 561-564.

[GOT78] Goto, S. and E. Kuh, An Approach to the Two-Dimensional Placement Problem in Circuit Layout, *IEEE Trans. Circuits and Syst.*, CAS-25(1978)218.

[GREE84] Greene, J.W. and K.J. Supowit, Simulated Annealing without Rejected Moves, *Proc. IEEE Int. Conference on Computer Design*, Port Chester, November 1984, pp. 658-663.

[GREE86] Greene, J.W. and K.J. Supowit, Simulated Annealing Without Rejected Moves, *IEEE Trans. on Computer-Aided Design*, CAD-5(1986)221-228.

[GRES86] Grest, G.S., C.M. Soukoulis and K. Levin, Cooling-Rate Dependence for the Spin-Glass Ground-State Energy: Implications for Optimization by Simulated Annealing, *Phys. Rev. Lett.*, 56(1986)1148-1151.

[GRÖ77] Grötschel, M., *Polyedrische Charakterisierungen Kombinatorischer Optimierungsprobleme* (in German), (Hain, Meisenheim am Glan, 1977).

[GRÖ84] Grötschel, M., Polyedrische Kombinatorik und Schnittebenenverfahren (in German), Universität Augsburg, *Preprint no. 38*, 1984.

[GRÖ87] Grötschel, M., *private communication*, 1987.

[GRO86] Grover, L.K., A New Simulated Annealing Algorithm for Standard Cell Placement, *Proc. IEEE Int. Conference on Computer-Aided Design*, Santa Clara, November 1986, pp. 378-380.

[GUL87] Güler, S., G. Garcia, L. Güler and M.N. Tokzöz, Detection of Geological Fault Lines in Radar Images, in: P.A. Devijver and J. Kittler, eds., *Pattern Recognition, Theory and Applications*, (Springer Verlag, Berlin, 1987), pp. 193-201.

[HAJ85] Hajek, B., A Tutorial Survey of Theory and Applications of Simulated Annealing, *Proc. 24th Conf. on Decision and Control*, Ft. Lauderdale, December 1985, pp. 755-760.

[HAJ86] Hajek, B., *private communication*, 1986.

[HAJ88] Hajek, B., Cooling Schedules for Optimal Annealing, *Mathematics of Operations Research*, 13(1988)311-329.

[HAS70] Hastings, W., Monte Carlo Sampling Methods using Markov Chains and Their Application, *Biometrika*, 57(1970)97-109.

[HIL85] Hillis, W.D., *The Connection Machine*, (MIT-Press, Cambridge, 1985).

[HIN83] Hinton, G.E. and T.J. Sejnowski, Optimal Perceptual Inference, *Proc. IEEE Conf. on Computer Vision and Pattern Recognition*, Washington DC, June 1983, pp. 448-453.

[HIN84] Hinton, G.E., T.J. Sejnowski and D.H. Ackley, Boltzmann Machines: Constraint Satisfaction Networks that Learn, Department of Computer Science, Carnegie-Mellon University, Pittsburgh, *Technical Report No. CMU-CS-84-119*, May 1984.

[HOP82] Hopfield, J.J., Neural Networks and Physical Systems with Emergent Collective Computational Abilities, *Proc. Nat. Academy of Sciences USA*, 79(1982)2554-2558.

[HOP85] Hopfield, J.J. and D.W. Tank, Neural Computation of Decisions in Optimization Problems, *Biological Cybernetics*, 52(1985)141-152.

[HUA86] Huang, M.D., F. Romeo and A.L. Sangiovanni-Vincentelli, An Efficient General Cooling Schedule for Simulated Annealing, *Proc. IEEE Int. Conference on Computer-Aided Design*, Santa Clara, November 1986, pp. 381-384.

[IRI83] Iri, M., K. Murota and S. Matsui, Heuristic for Planar Minimum-Weight Perfect Matching, *Network*, 13(1983)67-92.

[ISA74] Isaacson, D. and R. Madsen, Positive Columns for Stochastic Matrices, *J. Appl. Prob.*, 11(1974)829-834.

[ISA76] Isaacson, D. and R. Madsen, *Markov Chains*, (Wiley, New York, 1976).

[JEP83] Jepsen, D.W. and C.D. Gelatt Jr., Macro Placement by Monte Carlo Annealing, *Proc. IEEE Int. Conference on Computer Design*, Port Chester, November 1983, pp. 495-498.

[JES84] Jess, J.A.G., R.J. Jongen, P.A.C.M. Nuijten and J.C. Bu, A Gate Array Design System Adaptive to Many Technologies, *Proc. IEEE Int. Conference on Computer Design*, Port Chester, November 1984, pp. 338-343.

[JOHN85a] Johnson, D.S. and C.H. Papadimitriou, Computational Complexity, in: E.L. Lawler, J.K. Lenstra, A.H.G. Rinnooy Kan and D.B. Shmoys, eds., *The Traveling Salesman Problem*, (Wiley, Chichester, 1985), pp. 37-85.

[JOHN85b] Johnson, D.S. C.H. Papadimitriou and M. Yannakakis, How easy is local search?, *Proc. Annual Symp. of Foundations of Computer Science*, Los Angeles, 1985, pp. 39-42.

[JOHN87] Johnson, D.S., C.R. Aragon, L.A. McGeoch and C. Schevon, Optimization by Simulated Annealing: an Experimental Evaluation, Parts I and II, AT & T Bell Laboratories, *preprint*, 1987.

[JOHR82] Johri, A. and D.W. Matula, Probabilistic Bounds and Heuristic Algorithms for Coloring Large Random Graphs, *unpublished manuscript*, 1982.

[KAR72] Karp, R.M., Reducibility Among Combinatorial Problems, in: R.E. Miller and J.W. Thatcher, eds., *Complexity of Computer Computations*, (Plenum Press, New York, 1972), pp. 85-103.

[KER70] Kernighan, B.W. and S. Lin, An Efficient Heuristic Procedure for Partitioning Graphs, *Bell System Tech. J.*, 49(1970)291-307.

[KHA81] Khachaturyan, A., Semenovskaya, S. and B. Vainshtein, The Thermodynamical Approach to the Structure Analysis of Crystals, *Acta Cryst.*, A37(1981)742-754.

[KHA86] Khachaturyan, A., Statistical mechanics approach in minimizing a multivariable function, *J. Math. Phys.*, 27(1986)1834-1838.

[KIR82] Kirkpatrick, S., C.D. Gelatt Jr. and M.P. Vecchi, Optimization by Simulated Annealing, *IBM Research Report RC 9355*, 1982.

[KIR83] Kirkpatrick, S., C.D. Gelatt Jr. and M.P. Vecchi, Optimization by Simulated Annealing, *Science*, 220(1983)671-680.

[KIR84] Kirkpatrick, S., Optimization by Simulated annealing: Quantitative Studies, *J. Statis. Phys.*, 34(1984)975-986.

[KIR85] Kirkpatrick, S. and G. Toulouse, Configuration space analysis of travelling salesman problems, *J. Physique*, 46(1985)1277-1292.

[KOB84] Kobayashi, H. and C.E. Drozd, Efficient Algorithms for Routing Interchangeable Channels, *Proc. IEEE Int. Conference on Computer Design*, Port Chester, October 1984, pp. 79-81.

[KOZ62] Kozniewska, I., Ergodicité et stationnarité des chaînes de Markoff variables à un nombre fini d'états possibles, *Colloq. Math.*, 9(1962)333-346.

[KRA86a] Kravitz, S.A., Multiprocessor-Based Placement by Simulated Annealing, SRC-CMU Center for Computer-Aided Design, Department of Electrical and Computer Engineering, Carnegie-Mellon University, Pittsburgh, *Research Report No. CMUCAD-86-6*, February 1986.

[KRA86b] Kravitz, S.A. and R. Rutenbar, Multiprocessor-Based Placement by Simulated Annealing, *Proc. 23rd Des. Automation Conf.*, Las Vegas, June 1986, pp. 567-573.

[KRO71] Krolak, P.D., W. Felts and G. Marble, A Man-Machine Approach Toward Solving the Traveling Salesman Problem, *Comm. ACM*, 14(1971)327-334.

[LAA88a] Laarhoven, P.J.M. van, E.H.L. Aarts, J.H. van Lint and L.T. Wille, New Upper Bounds for the Football Pool Problem for 6, 7 and 8 Matches, to appear in: *J. Comb. Theory A*, 1988.

[LAA88b] Laarhoven, P.J.M. van, E.H.L. Aarts and J.K. Lenstra, Job Shop Scheduling by Simulated Annealing, Philips Research Laboratories, *preprint*, 1988.

[LAM86a] Lam, J. and J.-M. Delosme, Optimal Annealing Schedule, Yale University, New Haven, *Report 8608*, 1986.

[LAM86b] Lam, J. and J.-M. Delosme, Logic Minimization Using Simulated Annealing, *Proc. IEEE Int. Conference on Computer-Aided Design*, Santa Clara, November 1986, pp. 348-351.

[LAW85] Lawler, E.L., J.K. Lenstra, A.H.G. Rinnooy Kan and D.B. Shmoys, eds., *The Traveling Salesman Problem*, (Wiley, Chichester, 1985).

[LEN82] Lenstra, J.K., A.H.G. Rinnooy Kan and P. van Emde Boas, An Appraisal of Computational Complexity for Operation Researchers, *European J. of Oper. Res.*, 11(1982)201-210.

[LEO85a] Leong, H.W. and C.L. Liu, Permutation Channel Routing, *Proc. IEEE Int. Conference on Computer Design*, Port Chester, October 1985, pp. 579-584.

[LEO85b] Leong, H.W., D.F. Wong and C.L. Liu, A Simulated-Annealing Channel Router, *Proc. IEEE Int. Conference on Computer-Aided Design*, Santa Clara, November 1985, pp. 226-229.

[LEO86] Leong, H.W., A New Algorithm for Gate Matrix Layout, *Proc. IEEE Int. Conference on Computer-Aided Design*, Santa Clara, November 1986, pp. 316-319.

[LIG86] Ligthart, M.M., E.H.L. Aarts and F.P.M. Beenker, Design-for-testability of PLA's using Statistical Cooling, *Proc. 23rd Des. Automation Conf.*, Las Vegas, June 1986, pp. 339-345.

[LIN65] Lin, S., Computer Solutions of the Traveling Salesman Problem, *Bell System Tech. J.*, 44(1965)2245-2269.

[LIN73] Lin, S. and B.W. Kernighan, An Effective Heuristic Algorithm for the Traveling Salesman Problem, *Oper. Res.*, 21(1973)498-516.

[LINS84] Linsker, R., An iterative-Improvement penalty-function-driven wire routing system, *IBM J. Res. Developm.*, 28(1984)613-624.

[LUN85] Lundy, M., Applications of the annealing algorithm to combinatorial problems in statistics, *Biometrika*, 72(1985)191-198.

[LUN86] Lundy, M. and A. Mees, Convergence of an Annealing Algorithm, *Math. Prog.*, 34(1986)111-124.

[LUT86] Lutton, J.-L. and E. Bonomi, Simulated Annealing Algorithm for the Minimum Weighted Perfect Euclidean Matching Problem, *R.A.I.R.O. Recherche opérationnelle*, 20(1986)177-197.

[LYB85] Lyberatos, A., P. Wohlfarth and R.W. Chantrell, Simulated Annealing: An Application in Fine Particle Magnetism, *IEEE Trans. Magnetics*, MAG-21(1985)1277-1282.

[MAS85] Masarik, J., A Thermodynamically Motivated Optimization Algorithm: Circular Wheel Balance Optimization, *Aplikace Matematiky*, 20(1985)413-423.

[MET53] Metropolis, N., A. Rosenbluth, M. Rosenbluth, A. Teller and E. Teller, Equation of State Calculations by Fast Computing Machines, *J. of Chem. Physics*, 21(1953)1087-1092.

[MÉZ84] Mézard, M., G. Parisi, N. Sourlas, G. Toulouse and M. Virasoro, Nature of the Spin-Glass Phase, *Phys. Rev. Lett.*, 52(1984)1156-1159.

[MÉZ85] Mézard, M. and G. Parisi, Replicas and optimization, *J. Physique Lett.*, 46(1985)L-771 - L-778.

[MÉZ86] Mézard, M. and G. Parisi, A replica analysis of the travelling salesman problem, *J. Physique*, 47(1986)1285-1296.

[MÉZ87] Mézard, M., Spin Glasses and Optimization, in: J.L. van Hemmen and I. Morgenstern, eds., *Heidelberg Colloquium on Glassy Dynamics and Optimization*, *Springer Lecture Notes in Physics*, 275(1987)354-372.

[MIT86] Mitra, D., Romeo, F. and A.L. Sangiovanni-Vincentelli, Convergence and Finite-Time Behavior of Simulated Annealing, *Adv. Appl. Prob.*, 18(1986)747-771.

[MOO85] Moore, T.P. and A.J. de Geus, Simulated Annealing Controlled by a Rule-Based Expert System, *Proc. IEEE Int. Conference on Computer-Aided Design*, Santa Clara, November 1985, pp. 200-202.

[MORC86] Morgenstern, C.A. and H.D. Shapiro, Chromatic Number Approximation Using Simulated Annealing, Department of Computer Science, The University of New Mexico, Albuquerque, *Technical Report No. CS86-1*, 1986.

[MORI87] Morgenstern, I., Spin-glasses, Optimization and Neural Networks, in: J.L. van Hemmen and I. Morgenstern, eds., *Heidelberg Colloquium on Glassy Dynamics and Optimization*, *Springer Lecture Notes in Physics*, 275(1987)399-427.

[MOS86] Mosteller, R.C., Monte Carlo Methods for 2-D Compaction, Cal. Institute of Technology, Pasadena, *Ph.D. Dissertation*, 1986.

[NAH85] Nahar, S., S. Sahni and E. Shragowitz, Experiments with Simulated Annealing, *Proc. 22nd Des. Automation Conf.*, Las Vegas, June 1985, pp. 748-752.

[NAH86] Nahar, S., S. Sahni and E. Shragowitz, Simulated Annealing and Combinatorial Optimization, *Proc. 23rd Des. Automation Conf.*, Las Vegas, June 1986, pp. 293-299.

[NIC84] Nicholson, D.M., A. Chowdhary and L. Schwartz, Monte Carlo optimization of pair distribution functions: Application to the electronic structure of disordered metals, *Phys. Rev. B*, 29(1984)1633-1637.

[OSM87] Osman, W., Two-dimensional Compaction of Abstract Layouts with Statistical Cooling, *Philips Research Report*, 1987.

[OTT84] Otten, R.H.J.M. and L.P.P.P. van Ginneken, Floorplan Design using Simulated Annealing, *Proc. IEEE Int. Conference on Computer-Aided Design*, Santa Clara, November 1984, pp. 96-98.

[PAP77] Papadimitriou, C.H., The Probabilistic Analysis of Matching Heuristics, *Proc. 15th Ann. Allerton Conf. on Communication, Control and Computing*, 1977, pp. 368-378.

[PAP82] Papadimitriou, C.H. and K. Steiglitz, *Combinatorial Optimization: Algorithms and Complexity*, (Prentice-Hall, New York, 1982).

[PAR83] Parisi, G., Order Parameter for Spin-Glasses, *Phys. Rev. Lett.*, 50(1983)1946-1948.

[PIN70] Pincus, M., A Monte Carlo Method for the Approximate Solution of Certain Types of Constrained Optimization Problems, *Oper. Res.*, 18(1970)1225-1228.

[PIN86] Pincus, J.D. and A. Despain, Delay Reduction Using Simulated Annealing, *Proc. 23rd Des. Automation Conf.*, Las Vegas, June 1986, pp. 690-695.

[RAN86] Randelman, R.E. and G.S. Grest, N-City Traveling Salesman Problem: Optimization by Simulated Annealings, *J. Statis. Phys.*, 45(1986)885-890.

[RIN83] Rinnooy Kan, A.H.G. and G.T. Timmer, Stochastic Methods for Global Optimization, Econometric Institute, Erasmus University, Rotterdam, *Report 8317/0*, 1983.

[RIP86] Ripley, B.D., Statistics, Images and Pattern Recognition, *Canadian J. of Statistics*, 14(1986)83-102.

[RIV82] Rivest, R.L. and C.M. Fiduccia, A "Greedy" Channel Router, *Proc. 19th Design Automation Conference*, Las Vegas, June 1982, pp. 418-424.

[ROM84] Romeo, F., A.L. Sangiovanni-Vincentelli and C. Sechen, Research on Simulated Annealing at Berkeley, *Proc. IEEE Int. Conference on Computer Design*, Port Chester, November 1984, pp. 652-657.

[ROM85] Romeo, F. and A.L. Sangiovanni-Vincentelli, Probabilistic Hill Climbing Algorithms: Properties and Applications, *Proc. 1985 Chapel Hill Conference on VLSI*, May 1985, pp. 393-417.

[ROSE86] Rose, J.S., D.R. Blythe, W.M. Snelgrove and Z.G. Vranesic, Fast, High Quality VLSI Placement on an MIMD Multiprocessor, *Proc. IEEE Int. Conference on Computer-Aided Design*, Santa Clara, November 1986, pp. 42-45.

[ROSS86] Rossier, Y., M. Troyon and Th.M. Liebling, Probabilistic Exchange Algorithms and Euclidean Traveling Salesman Problems, *OR Spectrum*, 8(1986)151-164.

[ROT85] Rothman, D.H., Nonlinear inversion, statistical mechanics, and residual statics estimation, *Geophysics*, 50(1985)2784-2796.

[ROT86a] Rothman, D.H., Automatic estimation of large residual statics corrections, *Geophysics*, 51(1986)332-346.

[ROT86b] Rothman, S., Circuit Placement by Simulated Annealing at IBM, presented at the *Workshop on Statistical Physics in Engineering and Biology*, Lech, July 1986.

[ROW85a] Rowen, C. and J.L. Hennessy, Logic Minimization, Placement and Routing in SWAMI, *Proc. 1985 Custom IC Conf.*, Portland, May 1985.

[ROW85b] Rowen, C. and J.L. Hennessy, SWAMI: A Flexible Logic Implementation System, *Proc. 22nd Des. Automation Conf.*, Las Vegas, June 1985, pp. 169-175.

[RUT86] Rutenbar, R.A. and S.A. Kravitz, Layout by Annealing in a Parallel Environment, *Proc. IEEE Int. Conference on Computer Design*, Port Chester, October 1986, pp. 434-437.

[SAS88] Sasaki, G.H. and B. Hajek, The Time Complexity of Maximum Matching by Simulated Annealing, *Journal of the ACM*, 35(1988)387-403.

[SCH83] Schlag, M., Y.Z. Liao and C.K. Wong, An Algorithm for Optimal Two-Dimensional Compaction of VLSI Layouts, *Integration*, 1(1983) 179-209.

[SEC84] Sechen, C. and A.L. Sangiovanni-Vincentelli, The TimberWolf Placement and Routing Package, *Proc. 1984 Custom IC Conf.*, Rochester, May 1984, pp. 522-527.

[SEC85] Sechen, C. and A.L. Sangiovanni-Vincentelli, The TimberWolf Placement and Routing Package, *IEEE J. Solid State Circuits*, SC-20(1985)510-522.

[SEC86a] Sechen, C. and A.L. Sangiovanni-Vincentelli, TimberWolf3.2: A New Standard Cell Placement and Global Routing Package, *Proc. 23rd Des. Automation Conf.*, Las Vegas, June 1986, pp. 432-439.

[SEC86b] Sechen, C., Placement and Global Routing of Integrated Circuits Using the Simulated Annealing Algorithm, Univ. of California at Berkeley, *Ph.D. Dissertation*, 1986.

[SEM85] Semenovskaya, S., K.A. Khachaturyan and A.G. Khachaturyan, Statistical Mechanics Approach to the Structure Determination of a Crystal, *Acta Cryst.*, A41(1985)268-273.

[SEN81] Seneta, E., *Non-negative Matrices and Markov Chains*, (Springer Verlag, New York, 2nd ed., 1981).

[SHA48] Shannon, C.E., A Mathematical Theory of Communication, *Bell System Tech. J.*, 27(1948)379-623.

[SIA84] Siarry, P. and G. Dreyfus, An application of physical methods to the computer aided design of electronic circuits, *J. Physique Lett.*, 45(1984)L-39 - L-48.

[SIA85] Siarry, P., L. Bergonzi and G. Dreyfus, Optimisation du Placement de Blocs par la Méthode Thermodynamique: Application à la Conception du Plan de Masse d'un Circuit (in French), *Proc. Symposium on Custom Circuit Design*, Grenoble, 1985.

[SKI83] Skiscim, C.C. and B.L. Golden, Optimization by Simulated Annealing: A Preliminary Computational Study for the TSP, presented at the *N.I.H.E. Summer School on Combinatorial Optimization*, Dublin, 1983.

[SKR85] Skrien, D.J., An Algorithm for Optimal Multiple Foldings of Programmable Logic Arrays, University of Illinois at Urbana-Champaign, *M.Sc. Thesis*, 1985.

[SMI83] Smith, W.E., H.H. Barrett and R.G. Paxman, Reconstruction of Objects from Coded Images by Simulated Annealing, *Optics Letters*, 8(1983)199-201.

[SMI85] Smith, W.E., R.G. Paxman and H.H. Barrett, Application of Simulated Annealing to Coded-Aperture Design and Tomographic Reconstruction, *IEEE Trans. Nuclear Science*, NS-32(1985)758-761.

[SOL85] Solla, S., G. Sorkin and S. White, Configuration Space Analysis for Optimization Problems, in: E. Bienenstock, F. Fogelman Soulié and G. Weisbuch, eds., *Disordered Systems and Biological Organization*, (Springer Verlag, 1986), pp. 283-293.

[SON85] Sontag, E.D. and H.J. Sussmann, Image Restoration and Segmentation using the Annealing Algorithm, *Proc. 24th Conf. on Decision and Control*, Ft. Lauderdale, December 1985, pp. 768-773.

[SOU81] Soukup, J., Circuit Layout, *Proc. IEEE*, 69(1981)1281-1304.

[SPI85] Spira, P. and C. Hage, Hardware Acceleration of Gate Array Layout, *Proc. 22nd Des. Automation Conf.*, Las Vegas, June 1985, pp. 359-366.

[STE72] Stevens, J., Fast Heuristic Techniques for Placing and Wiring Printed Circuit Boards, University of Illinois at Urbana-Champaign, *Ph.D. Dissertation*, 1972.

[STO85] Storer, J.A., A.J. Nicas and J. Becker, Uniform Circuit Placement, in: P. Bertolazzi and F. Luccio, eds., *VLSI: Algorithms and Architectures*, (Elsevier's Science Publ., Amsterdam, 1985), pp. 255-273.

[SZU87] Szu, H. and R. Hartley, Fast Simulated Annealing, *Physics Letters A*, 122(1987)157-162.

[THO64] Thompson, G.L. and R.L. Karg, A Heuristic Approach to Solving Travelling Salesman Problems, *Management Sci.*, 10(1964)225-247.

[TOD83] Toda, M., R. Kubo and N. Saitô, *Statistical Physics*, (Springer Verlag, Berlin, 1983).

[TOU77] Toulouse, G., Theory of the frustration effect in spin glasses: I, *Communications on Physics*, 2(1977)115-119.

[UED83] Ueda K., T. Komatsubara and T. Hosaka, A Parallel Processing Approach for Logic Module Placement, *IEEE Trans. on Computer-Aided Design*, CAD-2(1983)39-47.

[UHR86] Uhry, J.-P., Simulated Annealing for Mesh Building and Other Problems, presented at the *Workshop on Statistical Physics in Engineering and Biology*, Lech, July 1986.

[VAND84] Vanderbilt, D. and S.G. Louie, A Monte Carlo Simulated Annealing Approach to Optimization over Continuous Variables, *J. Comput. Phys.*, 36(1984)259-271.

[VANN84] Vannimenus, J. and M. Mézard, On the statistical mechanics of optimization problems of the travelling salesman type, *J. Physique Lett.*, 45(1984)L-1145 - L-1153.

[VEC83] Vecchi, M.P. and S. Kirkpatrick, Global Wiring by Simulated Annealing, *IEEE Trans. on Computer-Aided Design*, CAD-2(1983)215-222.

[WAG75] Wagner, H.M., *Principles of Operations Research*, (Prentice-Hall, Englewood Cliffs, 1975).

[WEI67] Weinberger, A., Large Scale Integration of MOS Complex Logic: A Layout Method, *IEEE J. Solid State Circuits*, SC-2(1967)182-190.

[WHI84] White, S.R., Concepts of Scale in Simulated Annealing, *Proc. IEEE Int. Conference on Computer Design*, Port Chester, November 1984, pp. 646-651.

[WIL85] Wille, L.T. and J. Vennik, Electrostatic energy minimisation by simulated annealing, *J. Phys. A: Math. Gen.*, 18(1985)L1113-1117; Corrigendum, 19(1986)1983.

[WIL86a] Wille, L.T., Optimisation by Simulated Annealing: An Overview and some Case Studies, *CCP5 Inform. Q.*, 20(1986)19-31.

[WIL86b] Wille, L.T., Searching potential energy surfaces by simulated annealing, *Nature*, 324(1986)46-48.

[WIL87a] Wille, L.T., Minimum Energy Configurations of Atomic Clusters: New Results obtained by Simulated Annealing, *Chem. Phys. Lett.*, 133(1987)405-410.

[WIL87b] Wille, L.T., The Football Pool Problem for 6 Matches: A New Upper Bound Obtained by Simulated Annealing, *J. Comb. Theory A*, 45(1987)171-177.

[WOL85] Wolberg, G. and T. Pavlidis, Restoration of binary images using stochastic relaxation with annealing, *Pattern Recognition Letters*, 3(1985)375-388.

[WON86a] Wong, D.F., H.W. Leong and C.L. Liu, Multiple PLA Folding by the Method of Simulated Annealing, *Proc. 1986 Custom IC Conf.*, Rochester, May 1986, pp. 351-355.

[WON86b] Wong, D.F. and C.L. Liu, A New Algorithm for Floorplan Design, *Proc. 23rd Des. Automation Conf.*, Las Vegas, June 1986, pp. 101-107.

[WON88] Wong, D.F., H.W. Leong and C.L. Liu, *Simulated Annealing for VLSI Design*, (Kluwer, Boston, 1988).

[WOOD68] Wood, W.W., Monte Carlo Studies of Simple Liquid Models, in: H.N.V. Temperly, J.S. Rowlinson and G.S. Rushbrooke, eds., *Physics of Simple Liquids*, (North Holland, Amsterdam, 1968), pp. 117-230.

[WOOT85] Wooten, F., K. Winer and D. Weaire, Computer Generation of Structural Models of Amorphous Si and Ge, *Phys. Rev. Lett.*, 54(1985)1392-1395.

[WOU86] Wouters, C.R., E.H.L. Aarts and C.H. van Berkel, Optimization of Gate-Matrix Layouts using Simulated Annealing, *Philips Research Report*, 1986.

[YOS82] Yoshimura, T. and E.S. Kuh, Efficient Algorithms for Channel Routing, *IEEE Trans. on Computer-Aided Design*, CAD-1(1982)25-35.

[ZEE85] Zeestraten, R.J.A., Two-Dimensional Compaction, *Philips Research Report*, 1985.

Index

Aarts and Korst 25, 135, 147
Aarts and van Laarhoven 20, 22, 26, 39, 56, 60, 62, 65, 67, 70, 79, 95-97
Aarts *et al.* 47-48, 79, 95, 107, 140, 142
acceptance
  criterion 8-9, 150
  matrix (A) 13-15, 20-22, 24-27, 30-31, 33, 38, 57, 94
  probability 8, 13, 20, 23, 26, 45, 135
  ratio 59, 60, 135, 145, 150
Ackley *et al.* 25
algorithm
  auxiliary 145
  approximation 1-3, 6, 9, 55, 77, 87-88, 90-91, 99, 119, 126
  branch-and-bound 128
  convex hull 85
  deterministic 77
  exponential-time 2
  general 3
  homogeneous 14, 17-26, 55-56, 140
  inhomogeneous 14-15, 17, 27-38, 55-56, 81
  Kernighan-Lin 91
  learning 53, 133-135
  line restoration 131
  Lin-Kernighan 92
  Metropolis 8-9
  optimization 1, 3, 6, 55, 126, 128
  parallel 139-148
    (cf. parallel algorithms)
  polynomial-time 2, 90
  placement 101-102, 105, 108, 139
  probabilistic 77
  randomization 145
  routing 102, 116
  simplex 1
  tailored 3
  test 135
analysis
  average-case 77-78
  configuration space 40, 47, 52
  performance 77, 79, 82
    empirical 78-79, 95
    theoretical 78
  probabilistic value 78, 82-88
  worst-case 77-82
Anily and Federgruen 20, 22, 24, 28, 30-31, 37, 81
annealing of solids 5, 7, 39
approximation 6, 45, 48, 55-56, 99, 128, 135
Armstrong 135-136
aspect ratio 103
assignment problem 147

quadratic (QAP) 48, 82, 85-87, 98, 147
asymetric divergence 133-134
asymptotic convergence 6, 10-11, 14-15, 17-38, 52, 55, 82, 130, 151
average
cost 45-46, 65, 67, 70-71, 87
expected cost 45

balance
detailed 26
global 21
local 26
bandwidth 136
Banerjee and Jones 144-146
Bayesian approach 129
Beardwood *et al.* 50, 83
Beenker *et al.* 136-137
behavioural testing 118, 125
Bernasconi 39, 86, 136-137
binary sequence 86, 123-124, 137
biology 135-136
Biswas and Hamann 151
block 104-105
Bohachevsky *et al.* 150
Boolean
expression 120
function 118-120
operator 120
relation 132
Boltzmann
constant 7, 41
distribution 7-9, 41, 129-130
factor 8
machine 53, 131-135, 147
Bonomi and Lutton 39, 48, 60-61, 78, 82-86, 98
Bounds 40
bounding
box 103, 106
rectangle 101-102
Braun *et al.* 104
Burkard and Fincke 86
Burkard and Rendl 60-61, 98

Carnevalli *et al.* 50
carry output 118
Casotto *et al.* 144
Catthoor *et al.* 61, 118, 120
cavity method 40
cell 101, 105, 144
Černy 5, 61, 129, 150
Černy and Novák 147-148
channel 106, 110, 113-114, 116
Chung and Rao 144, 147
Chyan and Breuer 144
CIPAR package 105
code design 99, 136
combinatorial optimization (CO) 1-3, 6, 12
problems 6-7, 9, 39-40, 47-49, 50, 52, 78-80, 88, 99-100, 147
complexity
worst-case 4, 81
computer-aided circuit design 6, 99-128
condensed matter physics 7-8, 10, 39
conditions for convergence 19, 22-24, 26-27, 30-31, 33-34, 36-37, 56
necessary and sufficient 21, 25, 28, 35, 37-38
sufficient 24, 26, 28, 30, 32-33, 38
configuration 2-4, 8-9, 12-14, 17, 19, 21-24, 31-32, 35-36, 41, 43, 45, 52-53, 58, 60-

61, 63-67, 71, 73, 75, 78, 81, 100-128, 132-134, 141-142, 148-150
current 3, 9, 12, 14, 64, 75
density 45, 47-48
final 9, 66, 69, 80, 83, 88, 92, 127
globally minimal 10, 14, 18, 31-32, 55, 66, 77
initial 4-5, 10-11, 43, 94, 127, 141-142
locally maximal 32
minimum-energy 151
neighbouring 3, 10, 14
optimal 2, 5, 26, 43, 78, 80-81, 126, 136
set of 2, 4, 28, 33-34, 58, 80, 83, 85, 94
arbitrary 32, 38
space 2, 47, 52-53
connection
machine 131
strength 132-134
connectionist models 131
consensus function 132-133, 147
constraint
graph
vertical 114
horizontal 114-115
continuous-time analogon 34, 36
control parameter (c) 8-10, 13-15, 27-28, 42, 45-46, 49, 55-74, 79, 81, 84, 93, 95, 103-104, 130, 134, 136, 141, 144-146
decrement in 57, 61, 72, 84, 102
final value of 57, 60, 64-65, 71, 72, 79
initial value of 57-59, 62, 71-72, 90, 93
convergence 6, 12, 56
asymptotic 6, 10, 14, 17-38, 55, 82, 130, 151
in distribution 29
proof of 18
rate of 27, 34
cooling 7-8
rate 52, 95-96
schedule 57, 58-72, 73, 75, 77-79, 84, 88, 90-98, 100, 102, 105, 113, 116-117, 120, 123, 125-126, 141
conceptually simple 59-62, 71-72, 88, 95, 97, 117, 123
more elaborate 59, 62-72, 88, 96-97
cost function (C) 3-5, 8-10, 12, 26, 36, 44, 48, 50-53, 64, 66-67, 73-74, 90-91, 100-129, 148
errors in 105, 145-146
final value 87-88
count
tolerance 67
within 67
cup 34
depth of 35-36
crystal structure 151

Dantzig 1
Darema-Rogers *et al.* 144-146
decrement rule 57-58, 62, 67-72, 79, 81, 90, 98, 102, 150
delay 121
reduction 118
design automation 100-101
Deutsch 116
Devadas and Newton 109, 144-145

digital filter design 118, 120
direct problem 129
disordered metals 151
Distante and Piuri 118, 125
distribution
  Boltzmann 7-9, 41, 129-130
  Cauchy 151
  equilibrium 40-41
  Gibbs 41, 129
  normal (Gaussian) 47, 63, 67
  prior 130
  posterior 128, 130
  stationary 17-18, 20-21, 24, 30, 46, 56, 58, 67, 74-75, 87, 141
  uniform 9, 14, 18, 58, 78, 81, 83, 130, 149
don't care 123
Dunlop and Kernighan 109
dynamical windowing 145

efficiency 143-144, 146
eigenvalue 19, 29
eigenvector 19, 29, 30
El Gamal *et al.* 136-137
energy 7, 8, 40, 50, 52, 129
  average 12
  ground state 52
  Helmholtz free 42, 50
ensemble average 40
enthalpy 129
entropy 12, 40-44, 69-70, 85
  residual 44, 52
epoch 61
equilibrium 9, 85, 134
  distribution 40-41
  dynamics 40-44
  thermal 7-8, 40-42, 151
ergodicity
  coefficient of 29
  hypothesis 40
  theorem 28
Ettelaie and Moore 44, 52
evolutionary tree problem 136
exclusive OR 119
expected
  cost 41, 45-46, 48
  final cost 50, 95
  minimum cost 49
  squared cost 41

Feller 18
Felten *et al.* 144
Fleisher *et al.* 118-119
floorplan 121-123
floorplanning 118, 121-123
folding 118, 125-126
  multiple-column 125
  simple 125
  two-column 125
football pool 137
freezing 8-9, 49
frustration 52
Fu and Anderson 50
full adder 118
function
  Boolean 118-120
  configuration density 63
  consensus 132-133, 147
  cost (C) 3-5, 8-10, 12, 26, 36, 44, 48, 50-53, 64, 66-67, 73-74, 90-91, 100-129, 148
  distribution 151
  objective 150
  partition 7, 9, 41-42, 86
  penalty 102
  polynomial 1
  routing 107
  two-argument 21-22, 26

game theory 136
Garey and Johnson 1
gate array 107
gate matrix layout 108-109
Gelfand and Mitter 28, 32-33, 36-37, 81-82
Geman and Geman 28, 30-31, 37, 56, 128-131, 147
generation
  matrix (G) 13-15, 18, 20-22, 24-25, 27, 30-31, 33, 38, 57, 74
  mechanism 3, 4, 9, 59, 71, 73-75, 83-84, 100-128, 148-149
  probability 13, 135
Gibbs 40
Gidas 28, 34-35, 37-38, 40, 81, 128
Golay 86
Goldberg and Burstein 92
Golden and Skiscim 96
Gonsalves 118, 120
Goto and Kuh 106
graph 50-51, 80-81, 89, 91-94, 114, 120
  degree of 50
  problem
    coloring (GCP) 89-90, 92
    linear arrangement (GLAP) 93-94
    partitioning (GPP) 50-52, 75, 89-94, 97, 100, 109
  vertical constraint 114
Greene and Supowit 26, 73-75, 121
Grest *et al.* 39, 52
Grötschel 98
ground state 7, 43-44, 52
Grover 105
Güler *et al.* 131

Hajek 28, 34-38
Hamiltonian 50-53, 113
Hamming distance 124, 137
hardware acceleration 139
heat bath 7-8, 151
Hessian 149
heuristics 2, 91, 93
high temperature approximation 43
Hinton *et al.* 53, 131
Hopfield 52
Huang *et al.* 62, 65-67, 70-71, 105
hypercube 144, 147

image 129-131
  processing 51, 99, 128-131
  reconstruction 131
  restoration 147
  segmentation 131
integrated circuits 100
implementation
  micro-coded 139
  parallel 101, 139-148
  sequential 55-75
inverse problem 129
Isaacson and Madsen 29, 56
iterative improvement 3-5, 9-11, 52, 65, 98, 99, 108, 117-118, 120, 144-145

Jepsen and Gelatt Jr. 101
Jess *et al.* 107
Johnson *et al.* 59-61, 88-95
Johri and Matula 92

Kempe chain 92-93
Kernighan-Lin heuristics 91-92, 108-109
Khachaturyan 150
Kirkpatrick 88, 92

Kirkpatrick and Toulouse 40, 52
Kirkpatrick *et al.* 5, 39, 49-50, 52, 59-61, 83, 106
Kravitz and Rutenbar 144-146
Krolak *et al.* 96

Laarhoven, van *et al.* 96, 136, 137
Lam and Delosme 68, 97, 118-119
Langevin
  equation 40
  molecular dynamics 151
layout problem 101, 144
Lennard-Jones force 151
Leong 109
Leong and Liu 116
Leong *et al.* 60-61, 110, 113
Ligthart *et al.* 118, 123
limit cycle 120
Lin and Kernighan 83, 96
  2-opt 4, 83-85, 92
  3-opt 92
  k-opt 83, 92
linear
  arrangement problem
    graph (GLAP) 93-94
    min-cut 108
    net (NLAP) 93-94, 107-109
  array 107
linear programming 1, 120
Linsker 118
local search 3, 65
logic minimization 118-120, 144-145
  multi-level 119
longest-path problem 128
Lundy 135-136
Lundy and Mees 11, 21-23, 56, 60, 62, 64, 67-68, 79
Lutton and Bonomi 88
Lyberatos *et al.* 135

magnetics 135
magnetic field 136
  random 113
Markov chain 5-6, 12, 14-15, 21, 25-26, 45, 56, 60-61, 64-67, 70-72, 75, 82, 87, 91, 129-130, 140-143, 145-146, 149-150
  aperiodic 19, 24
  continuous-time 28, 36
  discrete-time 34, 36
  ergodic
    strongly 28-30, 34
    weakly 28-31
  homogeneous 13-14, 17-18, 20, 27, 29-30, 45, 56-57, 87, 130
  inhomogeneous 13-14, 27-29, 37
  irreducible 19, 24-25, 31, 36
  length 27, 56-58, 60-62, 65-67, 69, 71-75, 79, 84-85, 88, 90, 93, 95, 98, 102, 140, 142, 149
  sub-chain 140-143
Markov random field 129-130
Masarik 136-137
massive parallelism 131, 147
matching 80
  perfect 87
  problem
    minimum weighted (MWMP) 82, 87-88
    maximum 78, 80
materials science 136
matrix
  acceptance (A) 13-15, 20-22, 24-27, 30-31, 33, 38, 57, 94

covariance 149
distance 50, 82-83, 85, 87-88
generation (G) 13-15, 18, 20-22, 24-25, 27, 30-31, 33, 38, 57, 74
stochastic 13
transition 13-14, 19, 25, 27-28, 30, 36
maximum merit factor 86
maze router 113
mean 63, 67, 149
mesh
3-dimensional 137
Metropolis
algorithm 8-9
criterion 8-9, 14, 148
Metropolis *et al.* 8
Mézard 40
Mézard and Parisi 39-40
Mézard *et al.* 40, 52
micro
code 139
engine 139
minimum
global 4, 6, 12, 14, 21, 24, 27, 32-33, 35, 37-38, 52, 55, 80, 82, 87, 91, 96, 123, 128, 150
local 4, 5, 34-35, 52, 64, 66
Mitra *et al.* 28, 30-32, 37, 81
modules 101-107, 109-113, 121-123, 126, 128
Monte Carlo
annealing 5
experiments 83
method 8
Moore and de Geus 118, 125-126
Morgenstern, C.A. and Shapiro 60-61, 88, 92
Mosteller 118, 126
move class 73
multi-processor architecture 139

Nahar *et al.* 60-62, 88, 93
NAND 120
neighbourhood 3-4, 9, 14, 24, 61, 64, 75, 80, 83-84, 90, 149
pixel 129, 147
search 3
structure 9, 12, 83, 94
net 93, 101-103, 106, 107-117
net crossing histogram 106
net linear arrangement problem (NLAP) 93-94, 107-109
net partitioning problem (NePP) 75, 90-91, 107-108
neural networks 52, 99, 131, 147
Nicholson *et al.* 151
NOR 120
NP
complete 1-2, 52
hard 1, 3, 90, 100
number partitioning problem (NPP) 89
numerical analysis 135

optimization (CO) 1
combinatorial 1, 3, 6, 12
continuous 148-152
linear 1
non-linear 1
of cutting patterns 137
Osman 118, 126, 128
Otten and van Ginneken 21-22, 39, 60, 62, 65, 67, 69, 118, 121, 123

P 2, 78
Papadimitriou 88

parallel algorithms 139-148
 clustered 142-143
 general 140-143
 parallel moves 146
 partitioning 144
 single move decomposition 146
 systolic 140-143
 tailored 140, 143-147
Parisi 52
partition 89, 92-93, 109, 114-115, 125
 legal 92
 valid 115
partitioning problem
 graph (GPP) 50-52, 75, 89-94, 97, 100, 109
 net (NePP) 75, 90-91, 107-108
 number (NPP) 89
path 112
 L-shaped 112-113
 Z-shaped 112-113
penalty 103-105, 111, 121
performance 6, 11, 77-100, 106, 117, 120, 123, 126, 128, 144, 150-151
 in terms of computation time 4, 74, 96-98, 103, 105-106, 108-110, 116, 120, 123, 126, 128, 136, 139, 147, 150
 in terms of quality 6, 77-79, 85, 95-98, 103, 106, 108-109, 114, 116-117, 120, 123, 126, 128, 136, 143, 150
 in terms of running time 77-79, 85, 88, 95-96
permutation 82-83, 85, 87, 107, 109, 116, 125, 137
perturbation 3, 8-9
phase transition 40, 48-50
physics 39-53
Pincus and Despain 118, 121
pin 103, 116-117
pixel 129-130, 147
placement problem 48-49, 101-110, 121, 126, 144
 custom cell 102
 gate-array 105-106
 gate-matrix 108-109
 macro cell 102, 144
 standard cell 104-105, 109, 113, 144
point configuration 123
Polish expression 121-122
polynomial time 2, 78, 80, 96, 143
postulate 47-48
probabilistic
 exchange algorithm 5
 hill climbing 5, 14
 value analysis 78, 82-88
probability
 acceptance 8, 13, 20, 23, 26, 45, 135
 conditional 12-13
 density 45
  function 149
 distribution 8-9, 17, 56, 66-67, 77-78, 81, 129-130, 133, 135, 148, 150
  initial 17
 generation 13, 135
 transition 13-14
PROCEDURE SIMULATED ANNEALING 10
processor 140-148

stochastically coupled 148
Programmable Logic Array (PLA) 123, 125

quadratic assignment problem (QAP) 48, 82, 85-87, 98, 147
quality of final solution 6, 77-79, 85, 95-98, 103, 106, 108-109, 114, 116-117, 120, 123, 126, 128, 136, 143, 150
quantization mechanism 120
quasi-equilibrium 57-58, 61-62, 64-67, 70, 73, 140-141, 143
quenching 8, 10

Randelman and Grest 39, 79, 95
randomization
  algorithm 145
  technique 3
randomly generated problems 48, 50, 78, 84, 88, 91, 94-95, 97, 113, 116-117, 123
range limiter 103, 105
rate of convergence 27, 34
reachable at height 35-36
rejectionless method 74-75
renormalization group 129
replica analysis 40, 49-50
Ripley 131
Romeo and Sangiovanni-Vincentelli 20-21, 57, 66, 71, 97
Romeo *et al.* 21, 62
Rose *et al.* 144-145
Rossier *et al.* 25-26, 57, 98
Rothman D.H. 131
routing problem 51, 101-102, 110-118, 147
  channel 110, 113, 115, 117
    permutation 116
  global 110, 113, 144
Rowen and Hennessy 108

Sasaki and Hajek 80
Schlag *et al.* 128
Sechen and Sangiovanni-Vincentelli 60, 62, 101-102, 110, 113
seismology 131
Semenovskaya *et al.* 151
Seneta 28, 56
sequence
  of configurations 8-9, 35, 74
  of trials 12
  $\{c_k\}$ 15, 27, 30-31, 33-34, 38, 81
serializable subset 146
shape constraint 121, 123
Siarry *et al.* 103-104, 107
signal 109
silicon cluster 151
simulated annealing 3, 5, 7-15
Skiscim and Golden 60-62, 88, 94-95
slicing structure 121-122
solution 2-3
  final 2, 6, 77-79, 91, 94, 109
    quality 6, 77-79, 85, 95-98, 103, 106, 108-109, 114, 116-117, 120, 123, 126, 128, 136, 143, 150
  near-optimal 52, 98
  optimal 78, 85
specific heat 12, 42, 48-49
speed up 142
spin 51, 113
spin glass 40, 50-53
  Ising 44, 50, 53
standard deviation 46-48, 63

state 8
stationary distribution 17-18, 20-21, 24, 30, 46, 56, 58, 67, 74-75, 87, 141
statistical
  cooling 5
  mechanics 6, 8, 12, 39, 82
  physics 39-40
Steiner tree 136
Stevens 106
stochastic relaxation 5, 147
stochastically coupled processors 148
stop criterion 58, 60, 65, 79, 91
Storer *et al.* 107
system
  many-particle 40
  physical 43
  stochastic dynamical 66
  linear dynamical 66
Szu and Hartley 150

temperature 7-9, 13, 41
temperature regime 50
  high 50
  low 50
terminal 113-117
test-pattern generation 118, 123-125
thermodynamics
  third law of 43
TimberWolf package 102, 104-106, 113, 144, 147
time
  computation 4, 74, 96-98, 103, 105-106, 108-110, 116, 120, 123, 126, 128, 136, 139, 147, 150
  running 77-79, 85, 88, 95-96
timing 121
tour 2, 44, 82-84, 96, 144
  length 2, 50, 83-85, 95-98
transition 3, 5, 12-14, 17, 19, 27, 31-32, 34, 36-37, 55-61, 63-64, 66, 73-75, 78, 80-83, 85, 87-91, 96, 98, 100-128, 134, 140, 146, 149
  accepted 5, 26, 58-60, 65, 67, 71-75, 94, 142-143, 146
  class 73, 103
  matrix 13-14, 19, 25, 27-28, 30, 36
  probability 13-14
  rejected 61, 74, 93, 146
  2-opt 4, 83-85, 92
  3-opt 92
  k-opt 83, 92
tree
  Steiner 136
  evolutionary 136
trial 12-13, 45, 146
travelling salesman problem (TSP) 12, 43-44, 48, 52, 81-83, 85, 88-90, 92, 95-99, 143-144, 147
two-dimensional compaction problem 118, 126-128

Ueda *et al.* 144
Uhry 136-137
ultrametricity 40, 47, 50, 52-53
unit
  hidden 132
  input 132
  logical 132
  output 132

Vanderbilt and Louie 148-149
Vannimenus and Mézard 50
variable

binary 118-119
continuous 1, 148
discrete 1
variance 41, 45-46, 68, 149
vehicle routing 137
very large scale integration (VLSI) 110, 110, 125, 144
Vecchi and Kirkpatrick 50, 110, 112

weighting factor 51, 89, 116-117, 122, 124, 126
Weinberger array 108-109
wheel balance 136-137
White 39, 62-64, 73, 103
Wille 136-137, 152
Wille and Vennik 150
wire 101, 110, 114
length 103-104, 106-108, 123
Wolberg and Pavlidis 131
Wong and Liu 118, 121-123
Wong *et al.* 118, 125-126
Wooten *et al.* 136
Wouters *et al.* 109

Zeestraten 118, 126, 128